Structural Engineering Documents

12

Case Studies of Rehabilitation, Repair, Retrofitting, and Strengthening of Structures

International Association for Bridge and Structural Engineering IABSE
Association Internationale des Ponts et Charpentes AIPC
Internationale Vereinigung für Brückenbau und Hochbau IVBH

ISBN 978-3-85748-124-6

Publisher:
IABSE-AIPC-IVBH
ETH Hönggerberg
CH-8093 Zürich, Switzerland

Phone: Int. + 41-44-633 2647
Fax: Int. + 41-44-633 1241
E-mail: secretariat@iabse.org
Web: www.iabse.org

Preface

This monograph provides case studies of structural rehabilitation, repair, rehabilitation, strengthening, and upgrading of structures, which might be encompassed - in short - by the convenient umbrella terms "Conservation / Maintenance / Preservation / Upgrading of Existing Structures". Other umbrella terms for the activities related to maintaining and/or improving the structural performance of existing structures include: Restoration, Structural Renovation, Remedial actions. Rehabilitation and retrofit are sometimes used also as umbrella/generic terms. Terminology is discussed in more detail in the "Introduction".

The selected studies presented in this IABSE SED (Structural Engineering Document) cover a variety of structural types from different countries. SED 12 has been prepared as a joint activity between two IABSE Working Commissions WC3: *Concrete Structures*, and WC4 (formerly WC8): *Operation, Maintenance and Repair of Structures.*

A large part of existing buildings, bridges and other structures may have a long service life, where they could be subjected to severe environmental and/or operational conditions. These structures represent a strategic heritage of our societies and have an enormous economic value. Due to deficient or absent maintenance, changed operational conditions, new functional requirements, new code provisions, and/or safety necessities, a large number of structures could require to be structurally strengthened, repaired, upgraded, widened, refurbished, re-utilized, or rehabilitated. In most of the cases repair / modification is more convenient than replacement.

Strengthening, rehabilitation, repair, and retrofitting of structures is usually a challenging task because of uncertainties associated with old structures, restrictions on the geometry and materials used, and other structural or functional constraints.

When repairing / upgrading the structural performance of an existing structure, the engineers involved have plenty of possibilities, lots of constraints, and in some cases there are no applicable codes. Strengthening, rehabilitation, repair, and retrofitting is sometimes a complex and exciting work; an art. Restoration, structural renovation, and upgrading of structures is also involving enormous professional responsibility. This monograph is a summary of practices to help structural engineers. The reader will discover different approaches to put forward strengthening or rehabilitation projects. Even identical technical problems could have very different efficient solutions when considering the structural, environmental, and economic factors, as well as contractor and designer experience, materials, etc.

It was the initial intention for this SED to focus on concrete structures, with the hope that a future SED would cover other construction materials. Hence, the papers included in this monograph deal with Concrete Structures only. However, as the SED Editorial Board is planning for continuous addition of new case studies on conservation / upgrading of structures, through an electronic version of documents presented in this monograph, the SED Editorial Board requested that the scope of SED should be widened to cover other construction materials as well. Hence the title of the document was changed: "Concrete Structures" being replaced by "Structures".

The Editors have added four appendices which could provide information that might be of interest to the readers.

The Editors would like to thank the individual authors for their hard work and excellent papers. Special thanks go to the reviewers, Prof. Dr. Daia Zwicky and Mr. Joseph F. Tortorella, for their thorough reviews and useful comments. Thanks also to the Chair and Members of the IABSE SED Editorial Board for their cooperation in preparing the SED, and accepting its idea.

Mourad M. Bakhoum
Juan A. Sobrino

Table of Contents

Chapter

1 Introduction

1.1 Objectives and potential users

This document is intended as a guide for structural engineers and students. The objectives and potential users are outlined in the following:

(1) Many engineers are involved in the design and construction phases in the rehabilitation, repair, retrofit, strengthening, and upgrading of structures. These activities might be encompassed – in short- by one of the convenient umbrella terms: conservation/maintenance/preservation/upgrading of existing structures. A document presenting information, and discussing the different strategies and procedures considered in actual case studies from several countries, on a variety of conservation, preservation, restoration, structural renovation and upgrading projects would be of interest to and appreciated by engineers and students.

(2) Some universities offer courses (senior undergraduate and graduate level) on repair and maintenance of structures. Information presented in this monograph SED 12, should be useful to the students.

(3) Presenting information on repair procedures from different countries in a consistent format (section below) could provide a good example to the students and engineers on organizing the way they approach projects of repair/strengthening, and on writing reports about repair, strengthening, and upgrading of structures.

(4) Some of the repair/strengthening methods, such as external prestressing or glued laminates, although in use for many years, remain among the most commonly applied procedures on many structures; hence information presented in SED12 should be of value.

1.2 Format of the papers in SED 12

The papers in IABSE SED 12 are different from journal or conference papers in two main ways. Firstly, the number of pages of papers in SED 12 could be up to 25 pages, allowing the authors to provide more details than in journals or conference proceedings, where there are usually restrictions on the number of pages. Secondly, the contents of the papers follow a consistent format and organization of sections. This does not mean that it is the only possible format. There could be other formats, even better ones. However, providing a consistent format for all the papers is thought to be of good educational value – in particular for students. The following contents are proposed:

1. *Introduction*
2. *In-situ assessment and symptoms of distress, or reason for upgrading/widening..*
3. *Strategies for repair/strengthening/upgrading*
4. *Structural analysis before and after repair, design of sections, codes / guidelines*
5. *Detailing, and connections between new and existing parts of the structure*
6. *Construction procedures*
7. *Sustainability of repair/strengthening/upgrading procedures and environmental impact*
8. *Load testing / post intervention observations*
9. *Summary*
10. *References/Bibliography*
11. *Acknowledgements (if relevant)*

Note: Some deviations from this format were allowed in this 1st version of SED 12.

1.3 Importance and economic value of structural conservation, maintenance, preservation and upgrading activities

Conservation, maintenance, preservation and upgrading the structural performance of existing structures and infrastructures is of high importance, high economic return, hence funds are increasingly being devoted to these activities relevant to the existing structures. As an example, the following table shows the funds available for new structure works versus those for maintenance and repair works in seven countries [1, 2, 3, 4].

1.4 Additional information on structural conservation/maintenance/preservation/upgrading

Appendix A, B, C, D include additional information on Structural Conservation/ Maintenance/Preservation/Upgrading of existing structure.

Appendix A: List of Articles from IABSE – SEI Journals related to Topics of SED 12. Appendix A includes mainly a list of articles published in IABSE's journal *Structural Engineering International* (SEI), which are related to the topics of SED 12 (rehabilitation, repair, retrofit, strengthening, upgrading,. . ..of structures). These articles could present additional case studies to those presented in SED 12. The Appendix also includes information on where to order other IABSE publications related to topics of SED 12 (e.g. IABSE Conference Proceedings, and SEDs).

Appendix B: List of Some Codes, Guidelines, Manuals, Documents, and Books on Assessment, Conservation, Evaluation, Inspection, Maintenance, Preservation, Rehabilitation, Repair, Retrofit, Strengthening & Upgrading Structural Performance. Due to the importance, the large activities, and the high economic return of the structural conservation/maintenance/upgrading works, several books, codes, guidelines, online courses, reports, standards, and videos/presentations are currently available or have been recently published.

Country	New structure works	Maintenance and repair works	Total construction works
Japan*	52,5 trillion Yen (83%)	10,7 trillion Yen (17%)	63,2 trillion Yen (100%)
Korea*	116,8 trillion Won (85%)	21,1 trillion Won (15%)	137,9 trillion Won (100%)
France*	85,6 billion Euro (52%)	79,6 billion Euro (48%)	165,2 billion Euro (100%)
Germany*	99,7 billion Euro (50%)	99,0 billion Euro (50%)	198,7 billion Euro (100%)
Italy	58,6 billion Euro (43%)	76,8 billion Euro (57%)	135,4 billion Euro (100%)
UK*	60,7 billion Pounds (50%)	61,2 billion Pounds (50%)	121,9 billion Pounds (100%)
Switzerland**	29,1 billion Francs (62%)	17,9 billion Francs (38%)	47,0 billion Francs (100%)

Notes: (*) All the figures are for Year 2004, *except for Italy Year 2003.* (**) Year 2009

Table 1.1: Maintenance and repair works in different countries

Appendix B presents a partial list of these references. Appendix B lists also several references on "Terminology" relevant to topics of SED 12, as well as several references on "Checklists". Checklists could be a convenient tool for quality assurance and quality control of inspection conservation/maintenance/preservation/upgrading works.

Appendix C: Examples of Diagnostics of Crack Patterns & Causes of Deterioration in Concrete Structures. Appendix C includes examples of diagnostics of crack patterns and causes of deterioration in concrete structures. It is thought that these diagrams/tables of crack patterns and deterioration symptoms could be of high educational value, in particular to young engineers and students. The appendix includes few examples as a demonstration, and lists more than 40 references and websites for further information.

Appendix D: Guidelines on Selection of Rehabilitation, Repair, Retrofit Methods. The decision on which conservation/maintenance/preservation/upgrading procedures to use, which strengthening repair materials to select, which construction procedure to implement in the structural intervention is affected by many parameters, many constraints, making the process quite complicated. Appendix D presents examples for guidelines on selection of these procedures and methods from international codes, manuals, and recommendations.

Note: the websites mentioned in the Appendices were active in January 2010. They might change without notice.

1.5 Terminology and definition

This document provides case studies of structural rehabilitation, repair, retrofitting, strengthening, and upgrading of structures. It is quite interesting that different nouns and verbs are used in different countries, and in the different references to describe these activities, which are aiming at maintaining, restoring, improving, and/or upgrading the structural performance of existing structures. Moreover, different expressions are used for "umbrella terms" describing these activities, as mentioned in the Preface.

Appendix A presents a tentative grouping of words and verbs related to the Conservation/ Maintenance/Preservation/Upgrading of Structures. **Group (A):** This group is related mainly to the policies and planning of the activities in the post-construction phase, aiming at maintaining and/or improving structural performance. **Group (B):** This group is related mainly to observations and investigations carried out on the structures. **Group (C):** This group is related mainly to condition assessment and evaluation of structural performance. **Group (D):** This group is related to deterioration in materials and/or structures, which may result in a reduction of structural performance. **Group (E):** This group is related mainly to activities relevant to preventive maintenance. **Group (F):** This group is related to the changes in the dimensions of structural members (large structural intervention) to restore and/or upgrade the structural performance. Group (G) replacement and rebuilding of structures, Group (H): additional.

Appendix B lists several references on "Terminology" relevant to topics of SED 12. A brief overview of possible "Umbrella / Generic Terms" is presented in the following paragraphs: Conservation, Maintenance, Preservation, Upgrading of Structures. Rehabilitation and Retrofit are also very commonly used also in seismic engineering of structures.

Conservation: In the structural concrete Model Code revision, currently undertaken by the *fédération internationale du béton (fib),* it is suggested that these activities be encompassed by the convenient umbrella term "Conservation of Structures". Conservation would also include inspection, condition assessment, and regular maintenance activities for structures.

On the other hand, BD 89/03 from UK defines Conservation as: Conservation. Conservation is an approach where there is something of historic or aesthetic merit to be kept, but there can be change, as long as new insertions are in keeping or enhance that which is existing. It is a living and developing situation. For instance, saddling the arch of an old stone bridge or strengthening an existing parapet would be conservation, as would adding contemporary lighting in sympathy with the original design. Re-using an old highway bridge for pedestrians or cyclists where it was inadequate for motor vehicles would be a good example of conservation. Within an overall conservation exercise on a bridge there might well be restoration or preservation of certain elements. The principles of conservation are outlined in 3.1.

Upgrading: Upgrading is defined as "modifications to an existing structure to improve its structural performance" in ISO 13822: Basis for design of structures – Assessment of existing structures. The document includes also following definitions: Assessment: set of activities performed in order to verify the reliability of an existing structure for future use. Inspection: on-site non-destructive examination to establish the present condition of the structure. Maintenance: routine intervention to preserve appropriate structural performance. Rehabilitation: work required to repair, and possibly upgrade, an existing structure. Repair: improve the condition of a structure by restoring or replacing existing components that have been damaged. ISO 13822 presents a hierarchy of the terms. ISO TC 71 / SC 7 "Maintenance and Repair of Concrete Structures" is drafting an umbrella code for maintenance & repair. Terminology in the draft is prepared in accordance with ISO 13822.

Maintenance: The Japan Society of Civil Engineers issued the "Standard Specifications for Concrete Structures – Maintenance part" in 2001, and the 2nd edition 2007. Maintenance categories are defined as follows: (A) Preventive Maintenance: Maintenance to prevent visible deterioration on the structure during the service life. (B) Corrective Maintenance: Maintenance in which appropriate counter measures are taken after deterioration of the structure has appeared. (C) Observational Maintenance: Maintenance carried out primarily on the basis of visual inspection to structures without any direct measure and permits deterioration of the structure in a certain extent, or ones in which direct inspection is difficult or practically impossible to be carried out, such as underground structures.

The Manual of Maintenance of Steel Bridge Structures: Planning, Design, Construction for Maintenance and Durability, Hanshin Expressway Corporation, March 1993, Japan, defined Maintenance as: "Maintenance is the generic term for all jobs performed on the structure during its service life, that is, all jobs related to inspection, assessment, repair, reinforcement, replacement, improvement, database generation and input, and feedback operations for building new structures".

The AASHTO Maintenance Manual, section 3.1 mentions: "Bridge maintenance has been defined as work performed to keep a facility in its current condition. However, bridge maintenance has a broader scope because maintenance includes all activity in a facility's life that does not require a redesign and development project; thus, some agencies properly include work often classified as bridge rehabilitation (intended to upgrade the bridge to a condition better than its existing condition) within the context of bridge maintenance".

Preservation: The Bridge preservation Association in the USA defines Preservation as: "activities performed on bridge elements or components that aim to prevent, delay, or reduce deterioration. Bridge preservation activities do not entail structural or operational improvements of an existing bridge asset beyond its originally designed capacity". www.bridgepreservationassociation.org Moreover, The Safe, Accountable, Flexible, Efficient Transportation Equity Act: A Legacy for Users (SAFETEA-LU) Technical Corrections Act, enacted in June 2008, changed the Federal Highway Bridge Replacement and Rehabilitation Program to the Highway Bridge Program and placed greater emphasis on the importance of proper, timely bridge preservation. Highway Bridge Program funds can now be used for replacement, rehabilitation, painting, performing systematic preventive maintenance, seismic retrofitting, or applying anti-icing or deicing treatments to eligible highway bridge projects. http://www.fhwa.dot.gov/publications/focus/09may/01.cfm. Due to its importance, and in order to present a unified terminology, the FHWA (USA) issued a special memorandum on "Pavement Preservation Definitions".

Rehabilitation: "Is an all-encompassing term that includes concepts of repair, retrofitting, strengthening and weakening that may minimize the vulnerability of building structures to earthquake loading, " Seismic retrofitting of steel and composite building structures", by L. Di Sarno and A.S. Elnashai , Mid-America Earthquake Center Report, CD Release 02 - 01, University of Illinois (UIUC), USA, 2002. Page 3, presented the following:

"The terminology used in earthquake engineering for seismic rehabilitation of existing structures is open to misinterpretation. Therefore, common terms used in this report are italicized and defined as below". In this report:

Conventional intervention: Includes the established methods of repair, such as concrete encasement, use of bracings, strengthening or weakening of connections. By contrast no conventional intervention refers to the use of novel metals, namely aluminum, stainless steel and shape memory alloys, and/or special devices, e.g., base isolation and dampers which significantly enhance the energy dissipation and hence reduce story drifts and shears.

Rehabilitation methodology: Requires thorough assessment based on detailed as-built data and nonlinear static analyses either static (pushovers) or dynamic (time histories). Therefore, the refined approach is the more complete approach and can be applied to all structures.

Rehabilitation objective: Is the selection of desired damage levels or loss (performance levels) for a specific seismic demand (hazard level). Indeed, the performance levels define the expected behavior of the building in terms of allowable damage state to structural and nonstructural components for an identified earthquake ground motion.

Repair: Is defined as the reinstatement of the original characteristics of a damaged section or member and is confined to dealing with the as-built system.

Strengthening: The term strengthening is defined as the number of interventions that may improve one or more seismic response parameters (stiffness, strength and ductility) as a function of the desired structural performance level. Furthermore, strengthening includes the addition of structural elements or the change of the structural system.

Weakening: Is an alternative scheme to upgrade existing structures; it consists of reducing the seismic demand in critical regions, e.g., beam-to-column connections.

Renovation: Renovation is defined as a generic term in the book entitled: "Structural Renovation of Buildings: Methods, Details, & Design Examples, by A. Newman, McGraw-Hill Professional Publishing, 2001. Page 1, mentioned the following: *"Philosophers have long recognized that a fruitful discourse requires agreement on the terms of discussion. Various "R words" are used in this book to describe building renovation activities; these words sound similar but refer to slightly different concepts. Since there is no universal agreement on the meaning of these terms, the following common definitions are used here. . .."* The terms defined include: Rehabilitation, Remodelling, Renovation, Repair, Restoration, Retrofit.

Retrofit: Retrofit is used in some countries mainly for seismic upgrading of bridges. In other countries it is used as an umbrella term for all repair and strengthening activities.

Additional references on Terminology relevant to topics of SED 12, from different countries, are presented in Appendix B, section X.

[1] Performance-based Standard Specifications for Maintenance and Repair of Concrete Structures in Japan", by T. Ueda, K. Takewaka, IABSE SEI, 2007, Vol. 4.

[2] Japan Federation of Construction Contractors (http://www.nikkenren.com/), Japan Civil Eng. Contractors Ass. (http://www.dokokyo.or.jp/) & Building Contractors Society, *Kensetsugyo Construction Industry Handbook,* in Japanese,2006, (*http://www.bcs.or.jp/*).

[3] Research Institute of Construction and Economy, *Summary of 55th Euroconstruct Conference, Budapest,* (http://www.rice.or.jp/), 2003 (in Japanese).

[4] Swiss Federal Statistical Office, Bauausgaben und -vorhaben nach Art der Arbeiten, Swiss Federal Statistical Office (http://www.bfs.admin.ch), 2009 (in German)

[5] JSCE Guidelines for Retrofit of Concrete Structures – Draft, Concrete Library International, No. 36, December 2000, Japan. Ch. 7: Selection of Retrofitting Method.

[6] EN 1504 – Products and systems for the protection and repair of concrete structures – Definitions – Requirements – Quality control and evaluation of conformity, European standard. Part 9: 2008 General principles for use of products and systems.

Schematic Presentation

Activities on Conservation, Maintenance, Preservation, Upgrading, Rehabilitation, renovation, Retrofit, and routine/preventive maintenance of structures might be presented schematically as in the following graph:

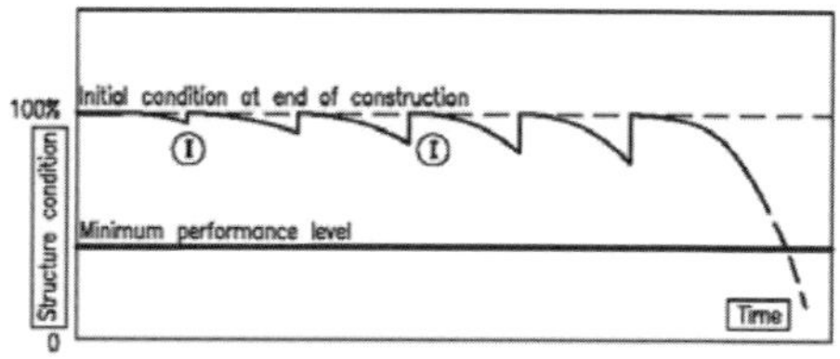

(i) At the time of the inspection, the structure fulfilled performance requirements but is predicted to not do so in the future due to a decline in performance due to load action and environmental action over time; performance improvements are conducted to prevent this in advance. Includes also routine maintenance activities such as painting of steel structures.

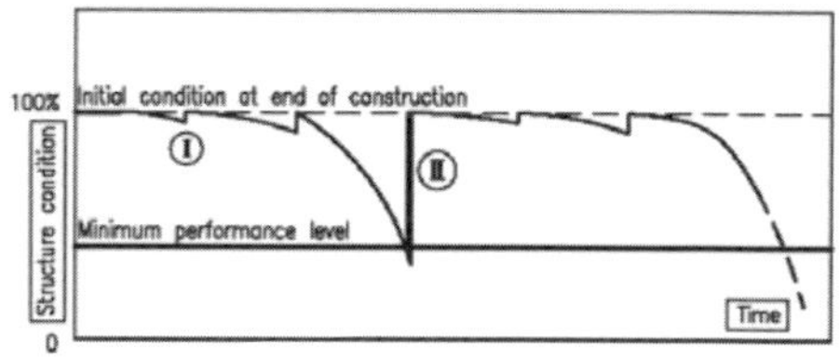

(ii) The performance requirements are the same as those of the structure when it was first built, but because the performance of the structure has declined due to load action and environmental action over time, the structure did not fulfill performance requirements at the time of the inspection; through retrofitting, the performance that would satisfy performance requirements is added.

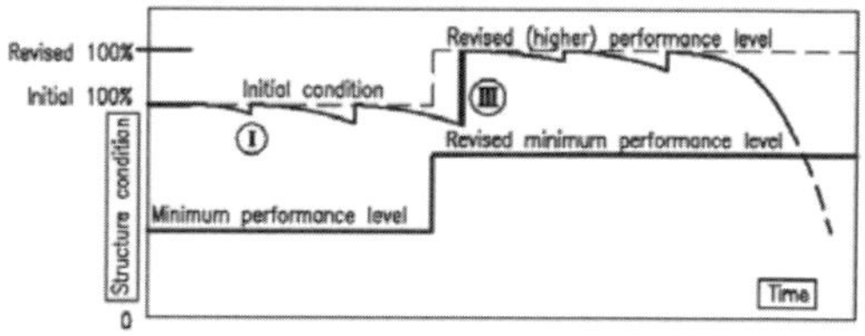

(iii) The design load has been changed or the structure otherwise requires a higher level of performance than when initially constructed, and therefore it does not fulfill performance requirements; through retrofitting, the performance that would satisfy performance requirements is added.

Quoted from JSCE Guidelines for Retrofit of Concrete Structures – Draft, Concrete Library International, No. 36, December 2000, Japan

(I) Regular maintenance (minor structural intervention) to maintain initial performance level.
(II) Remedial Actions (major structural intervention) such as Repair, Strengthening after major defect to restore the performance to initial performance level (initial condition at end of construction).
(III) Rehabilitation, Retrofit, Upgrading to a revised or higher performance level.

Fig. 1.1: Schematic Presentation of Typical Inspection/Maintenance/Repair/Upgrading cycles over the service life of a structure. (Based on: EN 1504-9:2008 Products and systems for protection and repair of concrete structures. Part9: Principles for the use of products and systems, Figure A.2, page 20)

Chapter

2

Upgrading the Seismic Safety of the Chritzi Bridge, Switzerland

Predrag Stefanovic, Structural Engineer, Emch + Berger SA Lausanne, Switzerland

Abstract: In the following text, a method of seismic safety improvement of bridges is proposed. It takes into account following requirements: structural security, serviceability, durability, and resistance towards earthquakes under conditions of the cost and value optimization.

Keywords: seismic safety improvement; elastic response spectrum; structural response.

2.1 Introduction

In the scope of the design methods, particularly of the dynamic loads, the engineer's knowledge and design codes have importantly advanced in the last decades. Most bridges of the Swiss national road system have been constructed before the introduction of the modern Seismic codes. The bridges, which have been built 30 and more years ago in the severe seismic Alpine regions, are to be examined within the maintenance and retrofitting processes. They are also examined for the seismic loads and have to be adjusted to the requirements of the presently valid design codes. This represents a challenging task for the design engineer.

Earthquake is a phenomenon of the rapid ground displacements with a general three-dimensional action vector towards the structure. The structure reacts to the excitation due to earthquakes in two ways: by the transfer and amplification of the ground displacements to its own structural body and by the generation of internal stresses within the structure. This pair of the interactive phenomena, displacements and stresses, is coupled and their relation in the structure is determined by the stiffness of the structural elements. The greater the stiffness of the bearing structure, the smaller are the deformations and the greater are the induced stresses and vice versa. Another problem arises from the spatial nature of the seismic propagation. As a result of wavelike earthquake oscillations, the pier foundations and abutments move asynchronously towards each other. The distance between different bridge supports increases and gets reduced periodically. Moreover, the ground settlement, liquefaction, instability, and collapse can result from an earthquake action.

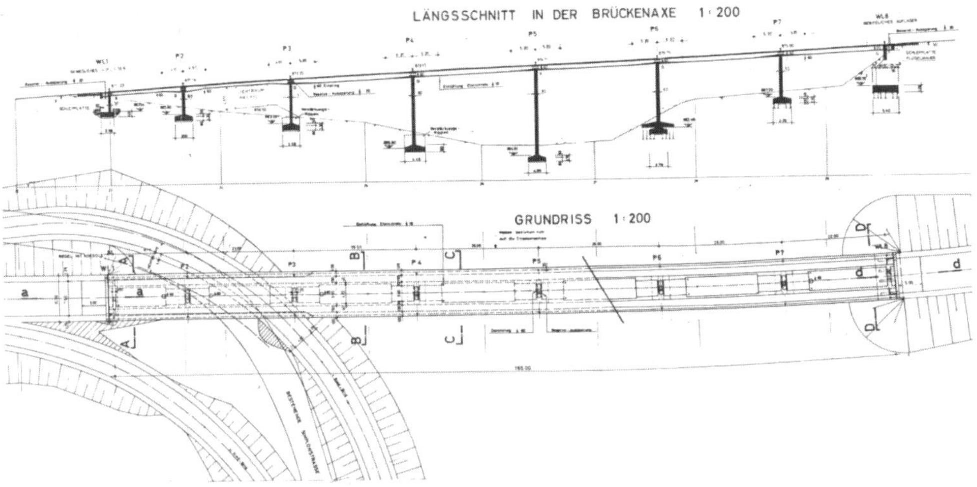

Fig. 2.1: Structure of the Chritzi Bridge before the repair [1]. Longitudinal section in the bridge axis 1:200 (scale); Situation 1:200

2.1.1 Project location, description of the structure, structural system

The bridge is situated close to the town of Brig on the Simplon Road connecting North Italy with the Swiss Canton of Valais (*Fig. 2.1*).

The Chritzi Bridge, which was submitted to the retrofitting project in 2001, is a post-tensioned concrete bridge of 165 m length. The seven-span (varying from 16 to 26 m) continuous beam with a cross section of a single cell box girder of constant height was built in 1971 conform to the SIA (Swiss) code for structural design valid at that time. The pillar heights range from 5.50 to 19.20 m, and their cross sections from 500/3000 to 800/3000 mm.

The location of the Chritzi Bridge belongs to the hazard zone Z3b (approximately eighth degree of Mercali Cancani Sieberg) after the design code SIA 160, 1993, valid in 2001 and is assigned to the structural class II as a bridge of importance for the accessibility to a region without acceptable alternative routes. The existing bridge was held for the horizontal forces in the longitudinal direction at the downward abutment (fixed bearings). In such a system, the structures' deformation is reduced to a minimum and consequently the seismic loadings due to an earthquake are at maximum.

Free sliding bearings, restrained in the transversal direction, were installed on the top of the pillar heads. The foundation conditions are regarded as rigid (rock). Foundation is partially mat slab foundation and partially on the concrete piles (*Fig. 2.2*).

2.2 Symptoms that led to need of repair/strengthening

A safety check of the existing bearing system, under the concurrent Swiss code of structural design (SIA 160 valid in 2001), has shown that fixed bearings at the abutments will collapse under the action of the higher seismic loads. After the collapse of fixed bearings, the

Fig. 2.2: Chritzi Bridge in its environment during the repair

bridge is not held anymore in the longitudinal direction and will be exposed to a series of shocks and non-repairable damages. In addition, the overall stability of the abutment could not have been guaranteed. This situation was regarded as not acceptable; thus, the seismic safety improvement ought to be performed.

2.3 Different strategies considered for seismic safety upgrading

Earthquake actions, as forces due to the structural stiffness, can be withstand by sufficient resistance and an elastic behaviour of the structure, or then they, as imposed deformations in form of ground displacements, can be withstand by sufficient ductility (deformation capacity) and the structure reacting in the plastic domain. The earthquake response of structures can be seen as a product of ductility and resistance. Historically, seismic safety improvement has been conventionally seen rather as a strengthening of the structural resistance. Conventional design was based on the resistance of the structure to withstand to earthquakes and maximal seismic loads in the elastic range of behaviour.

The upgrading of the seismic structural safety by conventional strengthening is usually quite difficult and costly and may not improve seismic resistance sufficiently. Moreover, it is not enough only to strengthen the vertical bearing elements, piers, and abutments, but their foundations should claim at least double safety against collapse compared with the superstructure. The rigid fixed bearings are constrained and their response may result in a brittle failure. Local damages are hardly controllable. The deflections of the swinging structure are reduced to a minimum, but the bridge can be stimulated to enter the higher oscillations forms. As the outcome of such an approach, the costs of the required strengthening can become essentially higher than the reduction of the seismic risk achieved.

Contrarily, an ideally elastic and ductile structure can be imagined. In this case, the induced internal stresses due to seismic loading are reduced to zero and no damages occur. The softer the structure, the lesser the sectional forces and stresses resulting from an earthquake. If the deflections are not limited, the serviceability might be compromised due to non-functional joints, high displacements paths of the bearings, and remaining deformations.

The bridge should be designed or retrofitted in a way that it sustains the ground displacements and oscillations due to an earthquake. The aim of the designing engineer is to provide sufficient seismic safety, to anticipate limited damages, and to take into account the cost optimization of the planned works. Damages can occur only in the limited intensity. Deflections sustainable for the serviceability are permitted by controlling the structural stiffness. With lesser stiffness, stresses due to seismic loading diminish. The principles of the seismic safe design state the following requirements:

a. Sufficient collapse safety and ductility: the bearing elements, pillars that take on the seismic loading, should be designed as ductile reinforced concrete walls.
b. Control of damages: to control the damaging effects of the earthquakes, it is necessary to provide a conceptual design of the structural deformations and the deformability of the non-carrying units (joints, canalization, equipment, cables). Finger joints have to be avoided. Joint design should provide for a free displacement of the bridge endings.
c. Clearly defined seismic behaviour, that is, structural response of the bridge structure: this should be achieved through a simple, systematic, and clear system for the transfer of the horizontal forces and the vertical loads. Such a system should provide a clear dynamic response of the structure and a predictable interaction with the ground.

For the determination of strategy, the following criteria have been considered: structural safety, serviceability, durability, feasibility of the realization, economic viability, deadlines and provisioning of the road traffic during construction. Several feasible strategies to counteract effects of earthquake have been compared and evaluated.

A. Structural strengthening in the conventional way was the first self-imposing idea. Within the existing bearing system, it should have been realized by raising the horizontal resistance of the bearing elements. New, more resistant shear bearings should have been built in. In addition, the overturning resistance of the abutment should have been increased by the additional post-tensioned ground anchors. An increase in the horizontal stiffness could also have been realized by the strengthening of the pillars. Strengthening can also unfavourably influence the seismic safety. As a result of the strengthening of resistance, the ductility is reduced and the stiffness is raised. The raised stiffness also means the higher natural frequencies, which generates greater seismic loads and sectional forces in the bridge structure and can lead in the extreme case to brittle failure.

B. Seismic isolation, that is, damping by the installation of the high-damping rubber bearings, hysteretic dampers or by a form of the hydraulic shock transmission device was another possibility. Seismic isolation of the structure provides reduction of the forces transmitted from the oscillating ground to the structure, without an increase of the stiffness. Simultaneously, an improvement of the collapsing safety (loss of support at the abutments and piers) by the change of geometry and the spatial arrangements of the bearing areas should have been performed (*Fig. 2.3*).

C. Shifting the natural frequency to a field of lesser acceleration response by the reduction of system stiffness has been analysed in detail. The horizontal structural stiffness determines

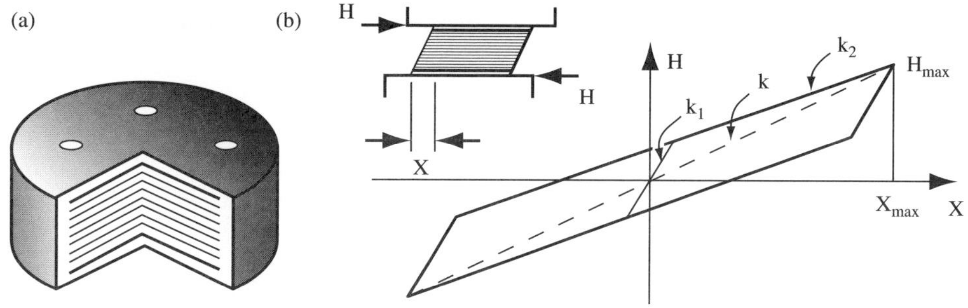

Fig. 2.3: Stratified seismic damping rubber bearing with idealized hysteretic curve [2]

dynamic response of the structure and the intensity of the earthquake loading. The natural frequency of the bridge is usually situated in the dropping part of the response spectrum diagram. The smaller the stiffness, the smaller the natural frequency of the structure and with it also the seismic loads, the sectional forces, and the required structural resistance. In this particular case of the Chritzi Bridge, the reduction of stiffness and of the natural frequency could have been realized by the modification of the structural system from "fixed" at one abutment to the "swimming" system in the longitudinal direction by releasing abutments and fixing the superstructure with the pillar heads. In this case, displacements due to earthquake become more important. With the serviceability becoming the main criterion of design, an appropriate enlargement of the abutment joints and deflection paths of the abutment bearings are imposed (*Fig. 2.4*).

Strategy C has manifested the lowest costs of the construction and maintenance, the shortest time of the realization, and reasonably simple construction method. Reasons for the selection of the strengthening method and materials are primarily in the simplicity and low cost of the execution, as well as the high durability of the selected structural behaviour and materials.

It was decided to modify the structural bearing of the bridge from the "fixed" into a "swimming" system. Through this transformation, entire structural system becomes "softer", the eigenvalues of the natural frequencies become lower and consequently the seismic loads as well.

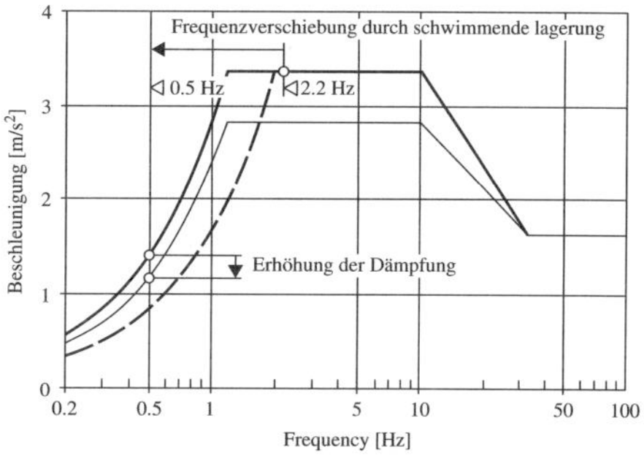

Fig. 2.4: Chritzi Bridge strategy C: shifting of the natural frequency to lesser acceleration [3]; Shifting of the natural frequency by the means of swimming supports; Increase of damping

2.4 Structural analysis before and after repair, design of sections, and codes

2.4.1 Structural analysis before and after repair

To modify the bridge structural system, fixed bearings at the abutment should have been replaced by the free sliding ones with a transversal restraint. It was decided not to replace

Risk case/loadings	Dead load	Traffic load	Brake forces	ΔT temperature
Earthquake	1.0/0.8	0.5 (M1)	—	—
Traffic loads	1.3/0.8	1.5	1.5	0.8

Table 2.1: Table of applied safety factors

the bearings at all piers (too costly and complex for the execution) but to fix the piers with the superstructure by the introduction of the shear bars beside the existing bearings. Basis for design is given by the following load combinations (code SIA 160, 1993) (*Table 2.1*).

2.4.1.1 Applied design methodology

To determine dynamic loading due to earthquake, the method of response spectrum was applied. Structural safety, as well as the global stability, has been checked with reduced pillar stiffness due to cracking. For the stability check of the pillars, the second-order effects have been considered. The bridge is held by the pillars in the longitudinal and transversal directions and across. Both bridge ends are free to move in the longitudinal sense. Seismic forces are transferred from the bridge superstructure via pillars to the foundations. Several successive approximations and iterations had to be performed.

First system approximation: "swimming bridge" with three middle piers fixed. Result: too large deformation, joints not feasible. Safety and serviceability in the exploitation are not provided.

Second system approximation: "swimming" bridge with four middle piers fixed. Result: still very large deformations and consequently costly joints.

Third system approximation: "swimming" bridge with all six piers fixed. Result: low displacements, economic joints. Structural safety and serviceability are provided.

First stiffness iteration: in the "swimming" system with all firmly connected piers and with full stiffness (non-cracked cross sections), natural frequencies and modal participation factors, and the intrinsic values of displacements for both horizontal (x, y) and vertical (z) directions have been determined. By means of the response spectrum analysis, all sectional forces (with $K = 2.5$ and $\mathrm{Cd} = 0.65$) were determined, as well as displacements (with $K = 1$ and $\mathrm{Cd} = 1$). Maximal longitudinal displacement of the non-cracked system amounts dx = 43.5 mm.

Second stiffness iteration: in the second iteration, the intrinsic values, as well as section forces and displacements, were calculated with the effective reduced pillar stiffness (from the M-EI_R interaction diagram). The structural safety of the pillars was checked by means of interaction diagrams for the sectional force combinations ($M_x/M_{zy}/N_z$) and has been approved as provided for the loading case of an earthquake.

Displacements and section forces due to earthquake have been determined in the system with reduced stiffness and cracked sections. Maximum abutment displacements amount:

- Displacement of the bridge structure ±73 mm due to earthquake;
- displacements due to asynchronous excitation of the abutments ±50 mm;
- maximum design movement 123 mm × 1.4 = ±160 mm determines the joint size.

As a critical load combination for the exploitation case stability check, a combination of the dead and traffic loads, braking forces, and temperature changes has also been considered. From the first calculation of the section forces, the effective stiffness is determined. In the cracked system with reduced stiffness, the effective displacements and section forces were determined. Maximum stretch amounts 73 mm. Structural safety and stability for both load situations including the second-order effects have been checked and approved.

2.4.2 Section analysis after repair

All cross sections of piers have been checked within the new structural system, after the repair. Structural safety checks have been performed for the cracked cross sections with the existing reinforcement. The bearing capacity check of all cross sections of piers, both in longitudinal and transversal directions, showed positive results. The existing concrete sections with their reinforcement piers disposed of sufficient resistance to withstand the maximal seismic loads and alternatively extreme combination of traffic loads, braking forces, temperature changes and shrinkage, creep imposed loads.

2.4.3 Design codes

Design code used in the repair project belongs to the SIA 160* family, particularly:

- SIA 161 (1989): Actions on Structures.
- SIA 160 (1970): Loading Conditions for Structures.
- SIA 162 (1989/1993): Concrete Structures.
- SIA 162 (1968): Reinforced and Post-tensioned Concrete Structures.
- SIA 162/5: Maintenance of Concrete Structures.
- SIA 462 (1994): Structural Safety of Existing Structures.
- SIA 469 (1997): Maintenance of Structures.
- Guideline ASTRA (1988): Observation and Maintenance of Structures of the Swiss National Roads.

SIA 160 design codes have been used in Switzerland until 2003. Presently, the Swiss code SIA 260 is valid, it is compatible with Euro codes.

2.5 Detailing

To materialize the new fixed joints on the pillar heads, the existing laterally restrained bearings have been kept. As the additional element aimed to the transition of the shear horizontal forces, a pair of shear dowels, corrosion-free steel bars have been selected ($D = 80$ mm, steel quality 1.4462) on the head of each pillar, connecting it stiffly with the bridge superstructure. These shear dowels have been introduced through the core-drilled recesses from the top of the bridge slab, aside the existing bearings. These recesses were subsequently filled in with special bearing mortar. The entire work could be executed in a very simple and rapid manner. Necessary structural modifications are presented in the following photographs (*Fig. 2.5*).

The inner space of the bridge cell box girder was basically occupied by the electric cables. To provide the security during the borehole works, these cables should have been located outside of the box girder.

Fig. 2.5: Rustproof steel bars D = 80 mm on the pier heads [1]

The replacement of the fixed abutment bearings by the longitudinally restrained new bearings required quite demanding works, while the road traffic had to be maintained during the entire operation. Because of special demands of the traffic, these works had to be carefully executed (*Figs. 2.6 and 2.7*).

For this purpose, the abutment body also had to be enlarged and reinforced. Part of the existing abutment was demolished and the reserves for new bearings were realized. To remove the existing fixed bearings from the abutment, the bridge had to be uplifted by the application of hydraulic jacks (*Figs. 2.8 and 2.9*).

2.6 Construction procedures

The possible strengthening measures have been limited by the specific project conditions and have influenced substantially the method of realization.

- Traffic on the Simplon Road, that is, over the bridge had to be maintained in every moment at least with one lane. By this condition, the step-by-step execution methods have been predefined.
- For the repair of the slab, the bridge has been entirely covered. This measure provided controlled climatic conditions, shelter against rain and wind, such that the required quality of the execution could have been realized (*Fig. 2.10*).
- The Chritzi Bridge is situated close to the residential area of Ried-Brig. To minimize the noise, the silent joints have been applied.

Fig. 2.6: Fixed bearings at the abutments were replaced by the unidirectional sliding bearings

Fig. 2.7: Uplifting of the bridge during replacement of the abutment bearings [1]

Fig. 2.8: Enlarging and reinforcement of the abutment walls [1]

Fig. 2.9: Joints at the bridge ends have been replaced by larger ones with V-shape [1]

Fig. 2.10: Covered building site of the Chritzi Bridge repair

2.7 Load testing

Load testing after finalizing the repair/strengthening of the Chritzi Bridge has not been performed. However, since its retrofitting in 2001, the bridge has withstood the series of minor earthquakes, which can be considered as the load testing.

2.8 Summary and conclusions

In the presented paper, a method of seismic safety improvement of bridges is proposed. It takes into account the following requirements: structural security, serviceability, durability, and resistance towards earthquakes under conditions of the cost and value optimization. Total intervention cost of 600 000 CHF represented around 15% of the cost of replacement by a new bridge.

The concept is simple and was realized with little expenditure with fully maintained traffic during the construction works, partly with traffic lights. The method and the implemented construction modifications can guarantee a durable resistance of the bridge for the horizontal seismic forces.

2.9 Acknowledgements

I wish to express my special thanks to Dr Thomas Wenk, expert of ASTRA, for his support in the presented project.

References

[1] BIAG Consulting Engineers ETH SIA USIC AG. *Project Documentation Erdbebensicherung Chritzibrücke*. BIAG Consulting Engineers ETH SIA USIC AG, Visp, 2001.

[2] Bachmann H. *Erdbebensicherung von Bauwerken*, 2 Auflage. Birkhäuser Verlag, Basel, 2002.

[3] Wenk T. *Beurteilung der Erdbebensicherheit bestehender Strassenbrücken*. Bundesamt für Strassen ASTRA, Bern, 2004.

[4] SIA 161. *Actions on Structures*. 1989.

[5] SIA 160. *Loading Conditions for Structures*. 1970.

[6] SIA 162. *Concrete Structures*. 1989/1993.

[7] SIA 162. *Reinforced and Post-Tensioned Concrete Structures*. 1968.

[8] SIA 162/5. *Maintenance of Concrete Structures*.

[9] SIA 462. *Structural Safety of Existing Structures*. 1994.

[10] SIA 469. *Maintenance of Structures*. 1997.

[11] Guideline ASTRA. *Observation and Maintenance of Structures of the Swiss National Roads*. 1988.

Chapter

3

Strengthening with Prestressed CFRP Strips of Box Girders on the Chofu Bridge, Japan

Masami Fujita, Manager, Yokkaichi Construction Office, Nagoya Branch, Central Nippon Expressway Co. Ltd., Yokkaichi, Mie Prefecture, Japan; **Terumitsu Takahashi**, Team Leader, Technical Department, DPS Bridge Works Co., Ltd., Toshima, Tokyo, Japan; **Kazuhiro Kuzume**, President, Kokusai Structural Engineering Corp., Nishi-ku, Osaka, Japan; **Tamon Ueda**, Professor, Division of Engineering and Policy for Sustainable Environment, Hokkaido University, Kita-ku, Sapporo, Japan and **Akira Kobayashi**, General Manager, Technical Development Department, Nippon Steel Composite Co., Ltd., Cyuo-ku, Tokyo, Japan

Abstract: Reinforced concrete (RC) box girders of the Chofu Bridge had been strengthened using tensioned carbon fibre reinforced polymer (CFRP) strip method. Before and after the CFRP application, on-site load tests of the bridge were conducted using a 45 t weight vehicle.

Keywords: tensioned CFRP strip; prestress; bending crack; deflection; natural frequency; strengthening.

3.1 Introduction

The Chofu Bridge of Chuo Highway is a three-span continuous reinforced concrete (RC) box girder bridge that was constructed 28 years ago and is located in the western part of Tokyo, Japan. The general view of the bridge is indicated in *Fig. 3.1*. The bridge condition had deteriorated through 28 years of heavy traffic loading and had many cracks on the underside of the main girders.

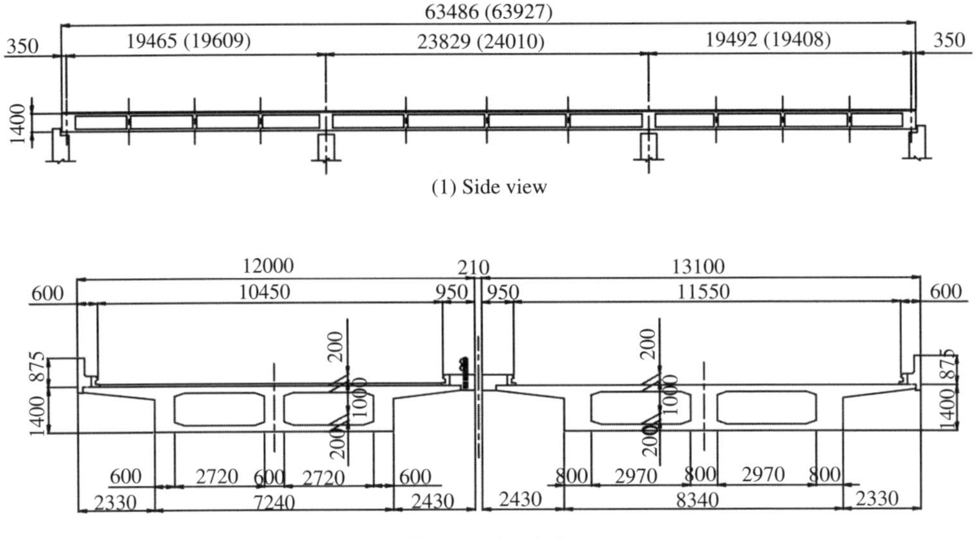

(1) Side view

(2) Cross sectional view

Fig. 3.1: General view of the Chofu Bridge (unit:mm)

3.2 *In situ* assessment and symptoms of distress

In the Chofu Bridge, which is an RC box girder bridge, many cracks had formed and water leakage from the cracks was observed at several locations (shown in *Figs. 3.2 and 3.3*) [1]. The deflection of the main girders caused by increase in heavy traffic and decrease in stiffness of the bridge also became obvious.

3.3 Strategies for strengthening

Steel plate bonding method and carbon fibre sheet bonding method were considered for this bridge. For the rehabilitation of this bridge, it was determined that passive applications such as steel plate or carbon fibre sheet bonding methods without prestressing were not enough. In order to increase the load carrying capacity and decrease the crack width, an external post-tensioning cable could have been used. However, in this case, restriction of the application spaces, flat surface terrain, and weak concrete, not suitable for attaching the brackets, made it impossible to apply external cables. On-site application without suspension of bridge service was also needed because the bridge has ramps of interchange and it was very difficult to stop the traffic.

At the Chofu Bridge, the tensioned carbon fibre reinforced polymer (CFRP) strip method was adopted because of various reasons; it is effective in reducing both dead load stress and dead load deflection, the prestress level of this method is suitable for the present condition, it is reasonably economical, and its application period is dramatically short. Passive applications such as steel plate or carbon fibre sheet bonding methods are not effective to reduce crack width or other effects due to dead loads, unless the structure is unloaded before application. One of the most advantageous points of the tensioned CFRP strip method is that by applying prestress to the concrete member, it is effective in increasing the structural performance of the

Fig. 3.2: State of damages

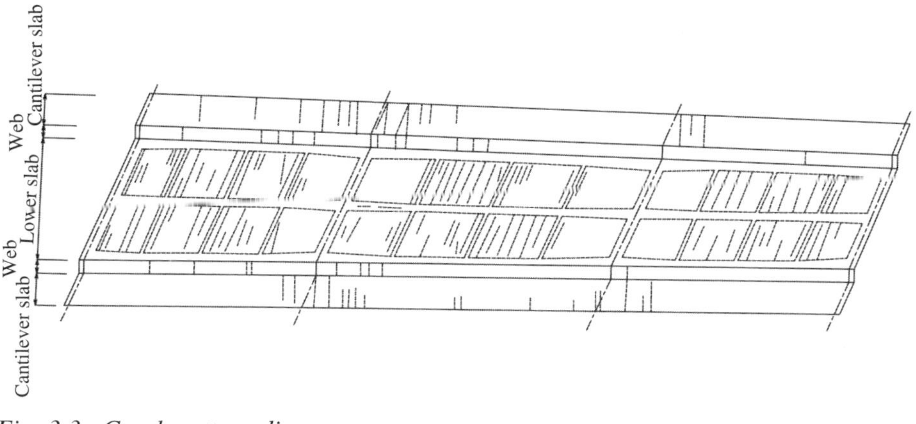

Fig. 3.3: Crack pattern diagram

member under not only live load but also dead load. Therefore, crack widths caused by dead load can be decreased by applying this method, which is very difficult when applying carbon fibre sheets passively. Another advantage is that by introducing a prestress, a redundant force is introduced to the upper surface of the continuous bridge girder, such that the tensile stress on the intermediate fulcrum is decreased. In other words, the upper surface of the intermediate fulcrum is also reinforced by this underside application. The distribution of bending moment is schematically shown in *Fig. 3.4* [2].

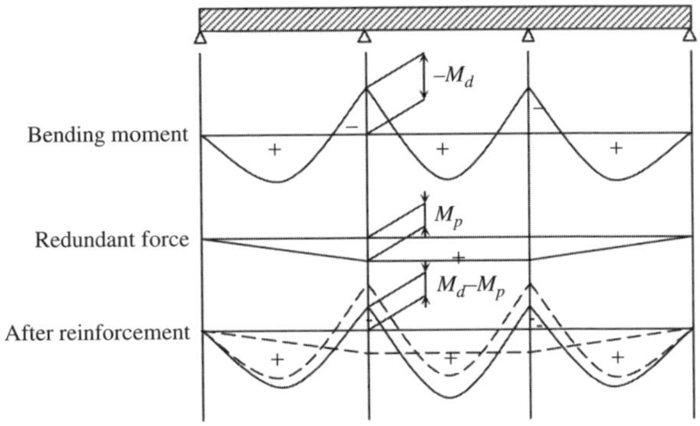

Fig. 3.4: Distribution of bending moment

The designed vehicle load had increased at the Chofu Bridge. In the strengthening design, with an aim of reducing induced stress under allowable stress, the structural safety was verified using allowable stress design method with consideration of external prestress provided by the tensioned CFRP strip according to specifications for highway bridges. A tension plate is made thin so that it can be applied to very narrow spaces. Because the introduced prestressing force is not too large (around 140 kN), the concrete stress at the anchoring area is also minimal. In the case of external cable post-tensioning, concrete stress at the anchoring area is sometimes so large that additional reinforcement at the anchoring area is required.

3.4 Structural analysis before and after repair, design of sections, and codes

3.4.1 Structural analysis before and after repair

The number of tension plates (CFRP strips) was determined by matching the allowable tensile stress in the longitudinal reinforcing bars considering the dead and live load bending moments, compressive stress, eccentric bending moment, and redundant force introduced by prestressing.

Here, 6% of design relaxation for CFRP strips was taken into account, whereas creep of concrete was not considered because it was assumed that creep of concrete had already converged after 28 years from construction. As a result, six CFRP strips were installed at the bottom of each RC box girder at each of the bridge spans. Only under the centre span of the bridge was the carbon fibre sheet method also applied because the tensioned CFRP strip method alone was not adequate to decrease tensile stress in the longitudinal reinforcing bar under design load. Allocation of CFRP strips and application procedure are shown in *Fig. 3.5*.

3.4.2 Codes

The codes applied for this rehabilitation work were as follows:

- *Design and Construction Manual for Outplate Method (Draft)*, Outplate Method Association, 2004 (in Japanese) [2].
- *Specification for Highway Bridge: Part 3 Concrete Bridge*, Japan Road Association, 2002 (in Japanese) [3].
- *Recommendation for Design and Construction of Concrete Structures Using Continuous Fiber Reinforcing Materials*, Japan Society of Civil Engineers, 1997 [4].

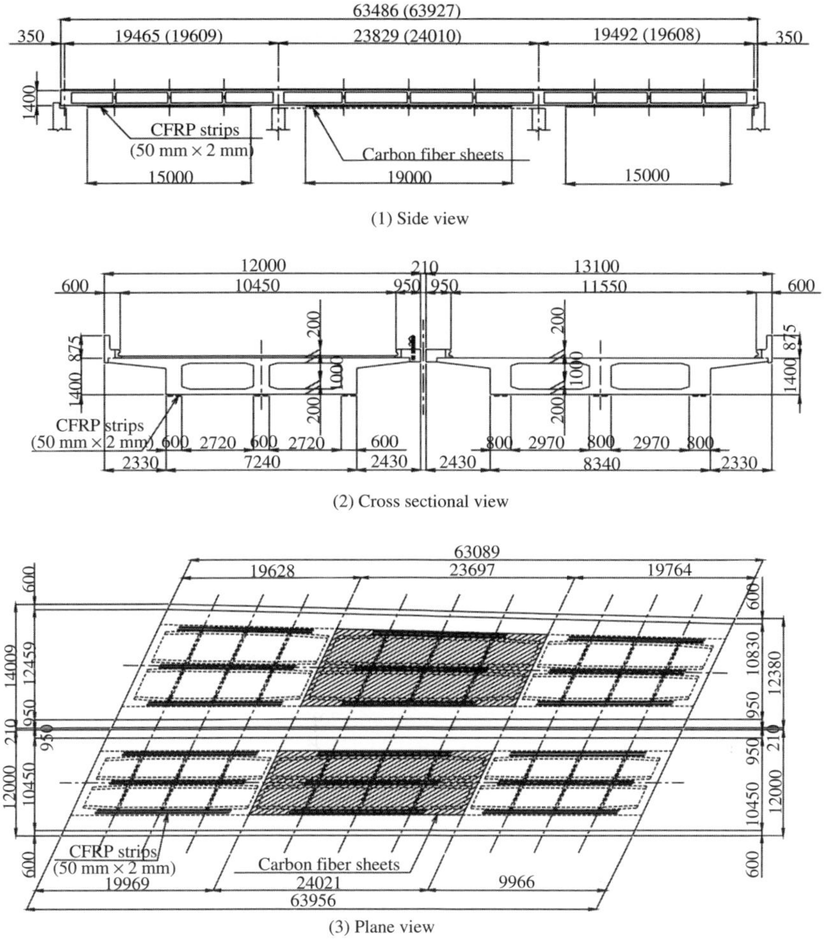

Fig. 3.5: Allocation of CFRP strips and CFRP sheets (unit:mm)

3.5 Detailing

The tensioned CFRP strip method consists of a tension plate, base plates, and intermediate anchoring devices. The tension plate is a CFRP strip that has anchoring devices at both ends. Base plates are steel frames that are attached to the concrete surface for anchoring the tension plate and for holding the specially modified hydraulic jack. Intermediate anchoring devices are simple stainless steel plates and anchor bolt (shown in *Fig. 3.6*). In order to prevent the CFRP strip from damage, a Teflon plate is inserted between the stainless plate and the CFRP strip. The CFRP strip is 2 mm thick and 50 mm wide. Each end of the strip is inserted into an anchoring device that is made of steel. Both ends are embedded and anchored by an expansive paste. This anchoring method is effective to reduce the stress concentration of CFRP strip at the anchoring device and full advantage of the high strength of CFRP can be taken [5]. The CFRP strip is made up of high strength carbon fibres and thermosetting resin, and fabricated using a pultrusion method. Characteristics of the CFRP strip are shown in *Table 3.1*. Any particular fire protection such as mortar coating was not used because in case of fire the

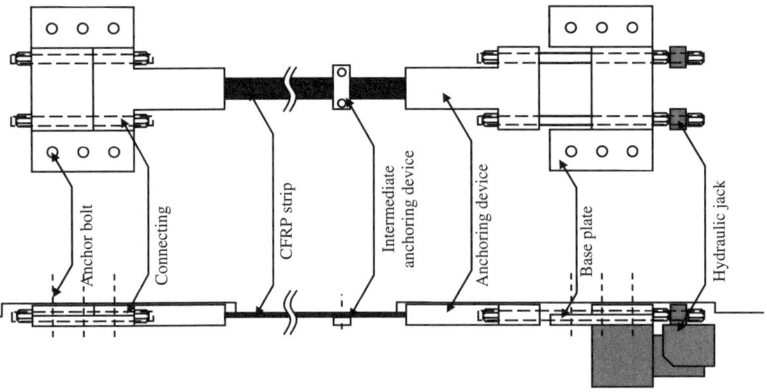

Fig. 3.6: Tensioning system

Width (mm)	50
Thickness (mm)	2
Tensile strength (N/mm^2)	2400
Modulus of elasticity (N/mm^2)	165 000
Tensile capacity (kN)	240
Producer	Nippon Steel Composite Co., Ltd.

Table 3.1: Characteristics of CFRP strip

concrete girders could bear the dead load without the CFRP strips and there was no risk of bridge collapse.

A hydraulic jack is specially designed for this application. It is divided into two parts (a reaction block and a cylinder) and each part is very light. Therefore, each part can be handled by one person. The weight of a base plate is light enough to be handled by two persons. This tensioning system does not require a reaction frame. The hydraulic jack is attached directly to the base plate so that the girder itself becomes the reaction frame.

3.6 Construction procedures

The application procedure is shown in *Fig. 3.7*. First, the concrete at both ends of the main girders is chipped away and the base plates are attached with anchor bolts. Second, the hydraulic jack, which is specially modified for this application, is set onto one of the base plates. One of the anchoring devices of the tension plate is bolted to the other base plate, and the other anchoring device is attached to the hydraulic jack.

Before tensioning the tension plate, adhesive resin is put on the upper surface of the CFRP strip. By tensioning the tension plate, compressive stress is introduced to the concrete girder. The CFRP strip is then adhered to the concrete surface. After tensioning, intermediate anchoring devices are attached with anchor bolts.

Using this procedure, tensioned CFRP strips are applied *in situ* to the underside of the main girders inside the bridge span as shown in *Fig. 3.8*.

3.7 Load testing

3.7.1 Outline of tests

A load test of the existing bridge was carried out to verify the strengthening effect. A test for prestress introduction was also conducted to examine the soundness of the concrete at the anchoring devices. Finite element (FE) analysis was done and results were compared with measured values.

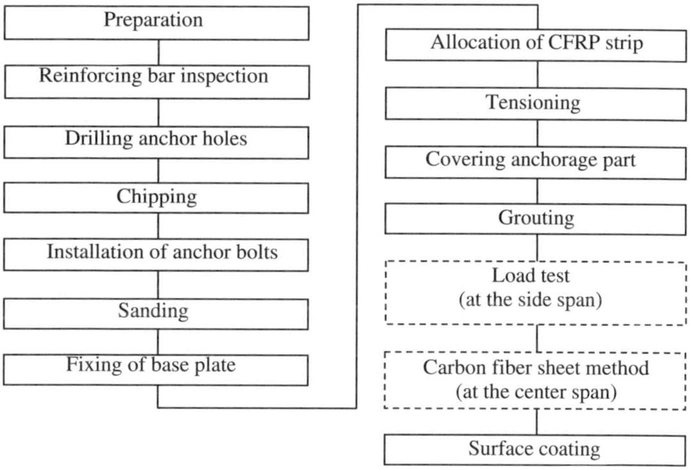

Fig. 3.7: Construction procedure

Fig. 3.8: Applied tensioned CFRP strip

3.7.1.1 Load testing

A 45 t weight vehicle (shown in *Fig. 3.9*) was placed statically at three different positions, as shown in *Fig. 3.10*. The stress in the longitudinal reinforcing bars, concrete crack width and depth, and deflection of the main girder were measured during the load testing, which was conducted both before and after the tensioned CFRP strip was applied. Crack width and depth were measured using displacement transducer and ultrasonic sensors, respectively. The 450 kN (45 t weight) load corresponds to 26.3% of the design live load at the centre of the side span. Measurement points are shown in *Fig. 3.11*.

Fig. 3.9: 45 t weight vehicle

In addition to the static load testing, dynamic load testing was also conducted by the 45 t weight vehicle driven at a speed of 50 km/h in the traffic lane, while all other traffic was stopped. The test was conducted at night. During the dynamic load test, the natural frequency of the main girder was measured with accelerometer.

3.7.1.2 Test for prestress introduction

The test for prestress introduction was carried out with the two CFRP strips applied to the centre web. Each CFRP strip was tensioned by a 160 kN force. The stress in the longitudinal reinforcing bars and concrete and crack widths in the concrete were measured at the centre of the span and at the vicinity of the anchoring devices. Measurement points are shown in *Fig. 3.12*.

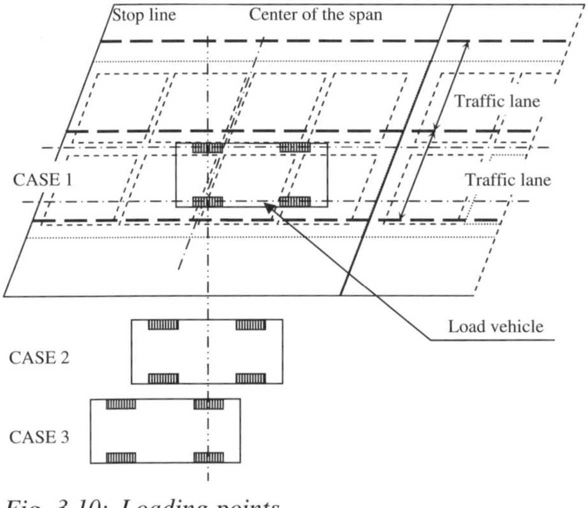

Fig. 3.10: Loading points

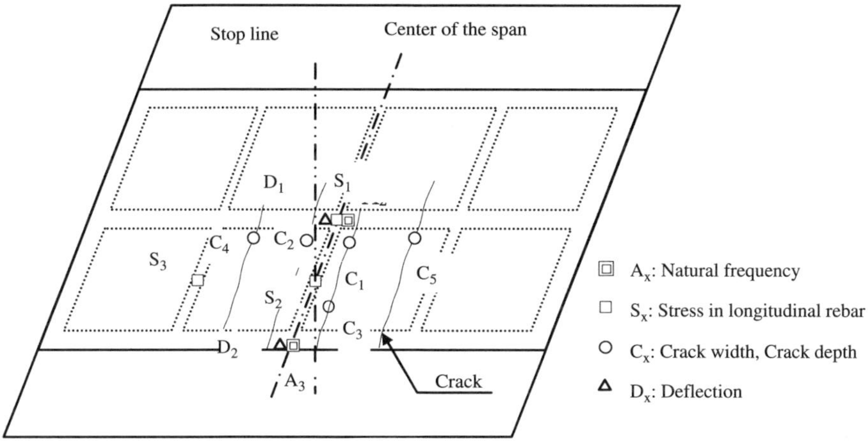

Fig. 3.11: Measurement points of load test

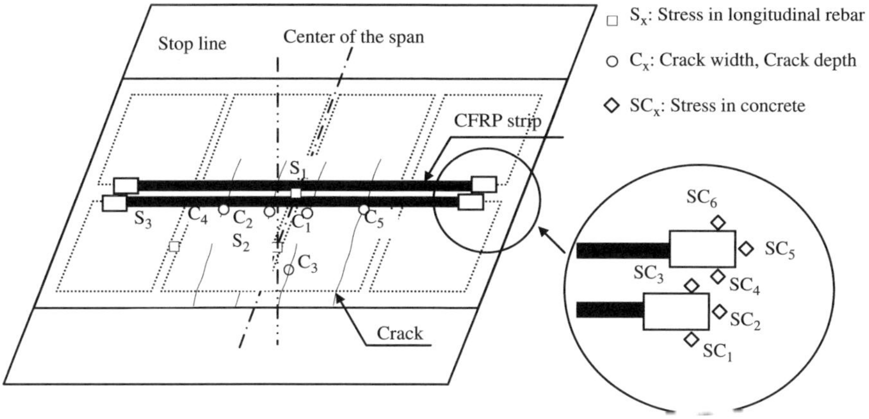

Fig. 3.12: Measurement points of prestress introducing test

3.7.2 Results and discussion

3.7.2.1 Load testing

3.7.2.1.1 Tensile stress in the longitudinal reinforcing bars. Observed tensile stresses in the longitudinal reinforcing bars under loading are shown in *Table 3.2*. Stresses in the longitudinal reinforcing bars were calculated by multiplying the measured strain with Young's modulus of 200 000 N/mm^2.

Because the subjected load of 450 kN (45 t weight) corresponds to only 26.3% of the design live load, measured stress of reinforcing bars was much smaller than allowable stress. At point S1 (centre of the span, centre in the direction of span width), the tensile stress before strengthening was 9.8 N/mm^2 and after the tensioned CFRP plate application was 7.8 N/mm^2. The latter tensile stress was 20% smaller than that of the former. It was confirmed that the

Point	Before strengthening (N/mm^2)	After strengthening (N/mm^2)
S1	9.8	7.8
S2	4.8	5.2
S3	6.6	5.0

Table 3.2: Observed tensile stresses in longitudinal reinforcing bars

Point	Before strengthening (mm)	After strengthening (mm)
C1	0.227	0.218
C2	0.226	0.188
C3	0.222	0.193
C4	0.107	0.090
C5	0.105	0.086

Table 3.3: Observed crack width

stress in the longitudinal reinforcing bar was decreased with the introduction of prestress by using the tensioned CFRP strip method.

3.7.2.1.2 Concrete crack width and depth. Observed concrete crack width and depth under loading are shown in *Tables 3.3 and 3.4*, respectively. Here crack depth was measured using ultrasonic sensors. At point C2 (centre of the span, centre in the direction of span width), the crack width decreased from 0.226 to 0.188 mm after strengthening (16.8% smaller) and also the crack depth decreased from 153 to 138 mm after strengthening (9.8% smaller). Average crack spacing around point C2 was about 1.0 m. It was confirmed that crack opening and extension were restricted with the introduction of prestress by using the tensioned CFRP strip method.

3.7.2.1.3 Deflection. Observed deflections under the loading are shown in *Table 3.5*. At point D1 (centre of the span, centre in the direction of span width), deflection after strengthening was decreased by 30% from 1.53 to 1.07 mm. Deflection was decreased by 40% from 1.88 to 1.12 mm at point D2 (centre of the span, edge side in the direction of

Point	Before strengthening (mm)	After strengthening (mm)
C1	130	102
C2	153	138
C3	112	99
C4	102	102
C5	162	146

Table 3.4: Observed crack depth

Point	Before strengthening (mm)	After strengthening (mm)	Strengthening effect ratio
D1	1.53	1.07	30.1% decrease
D2	1.88	1.12	40.4% decrease

Table 3.5: Observed deflection

Point	Before strengthening (Hz)	After strengthening (Hz)	Strengthening effect ratio	Analysis result (Hz)
A2	4.4	5.2	18.2% increase	5.393

Table 3.6: Natural frequency

span width). The deflections were measured as change of deflections under live load without effect of dead load, which mainly depend on stiffness of girder. Thus, it is clear that the stiffness of the main girder was improved dramatically.

3.7.2.1.4 Natural frequency. Observed natural frequency of the primary mode was improved from 4.4 Hz before strengthening to 5.2 Hz after strengthening. This increase in natural frequency corresponds to a mean decrease in amplitude, in other words, decrease in deflection. Therefore, this dynamic load testing result corresponds to the observed reduction in deflection during the static load testing (*Table 3.6*).

An eigenvalue analysis was conducted with a finite element method (FEM), assuming that there is no cracking in the concrete, as shown in *Fig. 3.13*. The natural frequency of the primary mode was determined to be 5.393 Hz, and this value was found to be very close to the measured value 5.2 Hz after strengthening application. This result implies that the rigidity of the main girder was improved by this strengthening method.

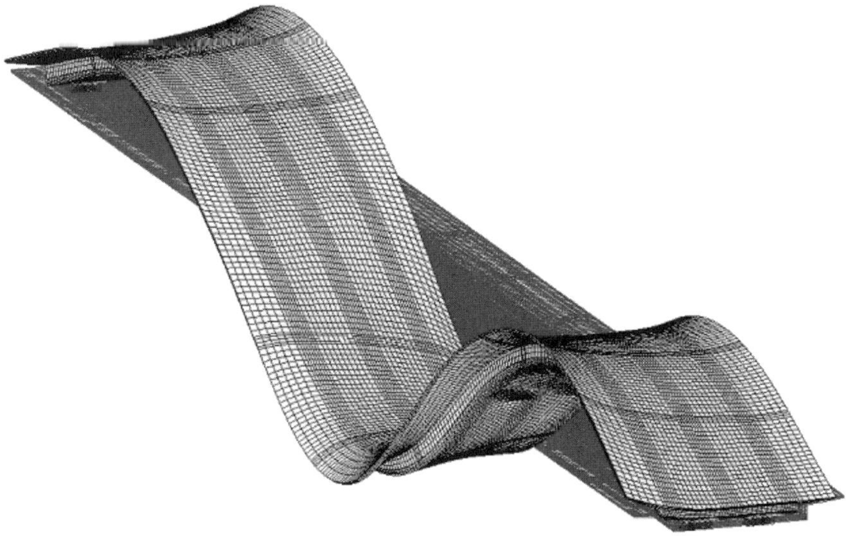

Fig. 3.13: Primary mode (FEM)

Point	Observed result (N/mm^2)
S1	2.4
S2	1.2
S3	1.8

Table 3.7: Compressive stress in longitudinal rebars

Point	Observed result
C1	0.007 mm decrease
C2	0.015 mm decrease
C3	0.012 mm decrease
C4	0.006 mm decrease

Table 3.8: Observed concrete crack width reduction

3.7.2.2 Test for prestress introduction

3.7.2.2.1 Compressive stress in the longitudinal reinforcing bar. In order to evaluate compressive stress that was introduced by tensioning CFRP strips, the change of stress in the longitudinal reinforcing bars was measured before and after prestressing. Observed results are shown in *Table 3.7*. The stress in the longitudinal reinforcing bar was calculated by multiplying measured strain by its Young's modulus of 200 000 N/mm^2. The compressive stress in the longitudinal reinforcing bar at the centre of the span (S1) was 2.4 N/mm^2 after tensioning two CFRP strips. This value agreed well with the 2.56 N/mm^2 stress calculated in the FEM analysis. Therefore, it was confirmed that the required prestress was introduced by tensioning two CFRP strips.

3.7.2.2.2 Concrete crack width. In order to evaluate change of crack width that was introduced by tensioning CFRP strips, the change of crack width at point C2 was measured before and after prestressing. Observed results are shown in *Table 3.8*. Concrete crack width at the centre of the span (C2) was decreased by 0.015 mm after tensioning two CFRP strips.

Point	Observed result (N/mm^2)
CS1	0.48
CS2	0.13
CS3	0.55
CS4	0.53
CS5	0.13
CS6	0.48

Table 3.9: Tensile stress in concrete

In order to verify this result, the crack width was calculated by using the crack width equation reported in Ref. [6]. Taking into account the effect of introducing prestress into the CFRP strip, the calculated crack width reduction was 0.02 mm. There was a good agreement between measured value and calculated value.

3.7.2.2.3 Tensile stress in concrete. The tensile stress in the concrete at the vicinity of the anchoring devices is shown in *Table 3.9*. Concrete stress was calculated by multiplying measured strain by its Young's modulus of 25 000 N/mm^2. After the CFRP strips were tensioned, tensile stress was caused in the concrete behind the anchoring devices. The maximum of this stress was 0.55 N/mm^2. Compared with the cracking limit of 1.9 N/mm^2 reported in Ref. [3], the measured value was small enough and very far from the stress that induces concrete cracking. Therefore, it was confirmed that the tensioned CFRP strip method is a safe method even for the concrete whose strength is relatively small when anchoring devices and bolts were installed.

3.8 Summary

Tensioned CFRP strip method was applied to the Chofu Bridge, a 28 year old RC box girder bridge, to rehabilitate and improve its structural performance. The cost for strengthening was about 15% of initial construction cost for the super structures. The condition of the Chofu Bridge had deteriorated by 28 years of heavy traffic loading and had many cracks on the underside of the main girders. Before and after the CFRP application, on-site load testing of the bridge was conducted using a 45 t weight vehicle. Results of the tensioned CFRP strip application to the bridge girders proved to be effective in reducing the stress in the reinforcing bars and in reducing crack widths. The most remarkable effects of this repair were a decrease in deflection at the centre of the span and an increase in the natural frequency of the primary mode. These effects imply that the stiffness of the main girder was improved.

References

[1] Fujita M, Imura M, Matsuda Y, Komada K, Takahashi T. Retrofit of Chofu Viaduct by out plate method and its load testing. *Bridge Foundation Eng.* 38(10): 15–19. 2004

[2] Outplate Method Association. *Design and Construction Manual for External-Plate Method (Draft)*. Outplate Method Association, Tokyo, 2007 (in Japanese).

[3] Japan Road Association. *Specification for Highway bridge: Part 3 Concrete Bridge*. Japan Road Association, Tokyo, 2002 (in Japanese).

[4] Japan Society of Civil Engineers. *Recommendation for Design and Construction of Concrete Structures Using Continuous Fiber Reinforcing Materials*. Japan Society of Civil Engineers, Tokyo, 1997.

[5] Hamada Y, Inoue M, Kobayashi A, Takagi N, Kojima T. Study on strengthening method for existing concrete structures using tensioned CFRP plate. Translation from *Proc. JSCE* 2002; No.711/V-56: 19–37.

[6] Japan Society of Civil Engineers. *Standard Specifications for Concrete Structures-2002 Structural Performance Verification*. Japan Society of Civil Engineers, Tokyo, 2002 (in Japanese).

Chapter

4

Punching Shear Strengthening at the New Station Square in Berne, Switzerland

Dominic Joray, Managing Engineer; **Martin Diggelmann**, Managing Director; Diggelmann + Partner AG, Berne, Switzerland

Abstract: The reinforced concrete slab of the reconstructed Station Square in Berne needed to be strengthened against punching shear. The case study led to the application of a newly developed post-installed punching shear reinforcement with inclined bonded bars.

Keywords: post-installed punching shear reinforcement; conservation; strengthening; inclined bonded bars; brittle failure; deformation capacity; construction process.

4.1 Introduction

The Station Square in Berne, Switzerland, as it is shown in *Fig. 4.1*, was constructed from 1971 to 1973 and reorganized and rehabilitated in 2007. The main element is an underground passage and shopping centre with an area of 7500 m^2. The ceiling is a 600 mm thick reinforced concrete slab that is mainly supported by steel columns. In front of the station building, a major city road, various tramways, and bus lines cross the square. The underground passage is about 134 m long and 42–61 m wide with a 54 m long and 16 m wide addition to the west. The clearance height is approximately 3.50 m. Several stairways and elevators around the perimeter give access to the underground passage.

There are 81 columns in total, usually in a grid of 8.44 m × 9.00 m. The columns are mainly steel pipes with an outer diameter of 368 mm and a thickness of 35 mm. Some columns consist of other steel profiles or cast-in-place reinforced concrete. The outer edge of the concrete slab is supported by reinforced concrete walls with neoprene bearings. The whole slab is divided into five elements. The initial design is based on the former Swiss codes SIA 160 (1970) [1] and SIA 162 (1968) [2]. The load model for traffic consisted of two axle loads of approximately 200 kN each and an accompanying load of approximately 5 kN/m^2 including a dynamic factor. The total dead load of road bed and pavement is about 30 kN/m^2.

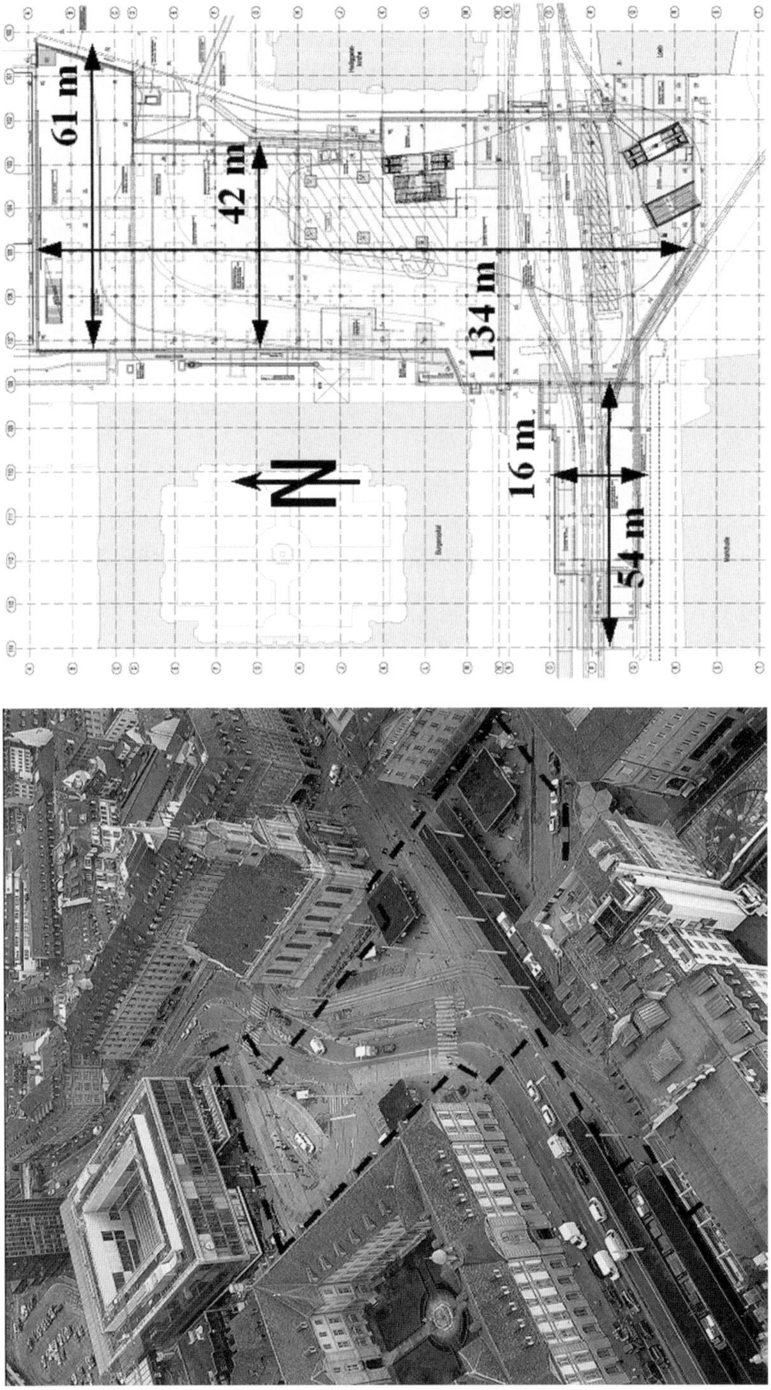

Fig. 4.1: View of the Station Square Berne before refurbishment and dimensions of the underground passage

In 2005, the tender was prepared including conservation and partial reconstruction of concrete members. The need for strengthening against punching shear according to the code generation introduced in Switzerland in 2003 was discovered already in the preliminary phase. In the further process, multiple strengthening methods to increase the punching shear resistance were investigated.

4.2 Symptoms that led to need of strengthening, and assessment of *in situ* conditions

4.2.1 Symptoms that led to need of strengthening

The reconstruction of the Station Square in Berne had to run through a political process including a public voting in the city of Berne. For quite a while, the necessity for repair and strengthening of several elements was identified. Because of the interdependent traffic systems and structural elements on the square, it was not reasonable to work on single elements though. Accordingly, a project complying with the requirements from public and individual transportation, local business, politics, functionality, and structural safety was developed. Apart from penetrating water at several locations, there were few of the usually expected symptoms such as rusting of steel reinforcement, spalling, or cracks. Nevertheless, the concrete slab had to be supported by temporary wood logs at a few locations.

The need for strengthening against punching shear was identified during the preparation of the tender documents in 2005. Shortly before, the new Swiss codes [3–5] were introduced in 2003 including more stringent requirements on punching shear resistance due to increased knowledge gained in the past years. The preliminary reconstruction project of the Station Square additionally led to an increase in load on the existing concrete slab.

4.2.2 Assessment of *in situ* conditions

In the tender documents, a condition survey was included. At this stage, the following investigations were conducted: a total of 12 openings were created in different concrete members to assess the actual condition; 22 core samples were taken to investigate compressive strength, chloride penetration, carbonation depth, and porosity. The average compressive strength of 19 specimens was 47.1 N/mm^2 with a minimum of 31.7 N/mm^2. The tested specimens were taken close to the slab edge. The waterproofing system was investigated at five openings. The reinforcement cover and the adhesive tensile strength were measured at a few locations as well.

For the construction work and the structural analysis, more tests were necessary. 14 more core samples of different lengths were taken, and at 14 additional locations, the adhesive tensile strength was measured. The reinforcement cover was examined all over the concrete slab. Because of the limited damage from chlorides and carbonation and the clear results from the tender documents, no additional tests were conducted. From the additional cores, 28 specimens were tested to determine a compressive strength value according to Ref. [6]. The tested specimens were mainly taken from the core of the slab. The average compressive strength including the results from the initial survey was determined as 61 N/mm^2. On the basis of Ref. [6], a concrete quality of C35/45 ($f_{ck}/f_{ck,cube}$) could be assumed for the structural analysis. The steel pipes and the reinforcement were tested as well to affirm the assumed design values.

4.3 Different strategies considered for repair

4.3.1 Considered solutions

4.3.1.1 Vertical post-installed punching shear reinforcement

The vertical post-installed punching shear reinforcement set in core drilled holes with anchor heads on both sides (*Fig. 4.2*) is a relatively simple and effective strengthening method. It can be designed like regular punching shear reinforcement. To avoid damaging, the tensile reinforcement on the upper side has to be marked. Because of the core drilled holes, damage to the lower tensile reinforcement is hardly avoidable. At least one side of the post-installed bars needs to have a thread to apply a nut. It would be possible to embed nut and washer in the concrete slab, although commonly they are applied onto the lower concrete surface. In addition to the increased punching shear resistance, this strengthening method also increases the deformation capacity of the flat slab.

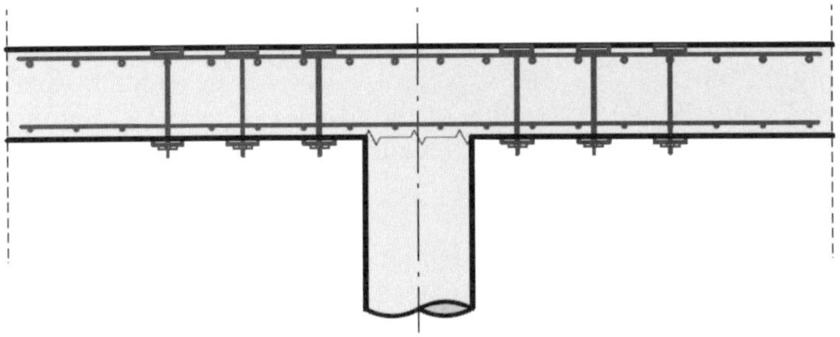

Fig. 4.2: Exemplary vertical post-installed punching shear reinforcement [7]

However, a simultaneous accessibility on both sides of the slab is necessary. Due to the complex construction process on the upper side because of car traffic, bus lines, and tramways, this requirement would have resulted in too many constraints. The time for construction on upper and lower sides individually was very limited already. Thus, the vertical post-installed punching shear reinforcement was not executed.

4.3.1.2 Steel collars

The tender documents suggested steel collars (*Fig. 4.3*) around the columns to increase the punching shear resistance. During the planning process, the details were adjusted several times. One goal was to reduce welding on site, especially at not verifiable locations. Another restriction was the limited total height of the collar of 350 mm requested by the client. All structural elements were requested to be situated above illuminating level.

The finally considered rectangular collar had a width of 1.3 m and a height of 330 mm. The total weight of one collar was 1300 kg. It consisted of two elements that had to be connected with prestressed bolts on site. Additionally, a steel support had to be welded to the columns in advance.

Despite the improvements, several details could not be solved to full satisfaction. Almost 60 collars would have to be produced and mounted. The weight of 650 kg per element would hinder the execution on site and the total amount of steel leads to inappropriate cost. Additionally, it was uncertain what amount of load could be carried at the expected small deformation due to the stiff behaviour of the concrete slab of 600 mm thickness on a span of only 8.44–9.00 m. Possibilities to prestress or preload the steel collars were evaluated to

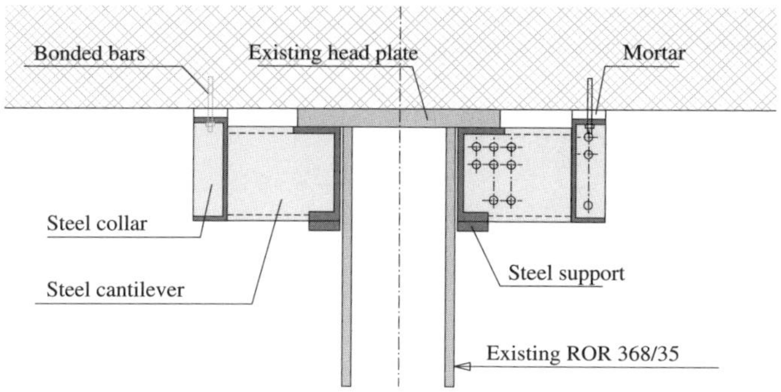

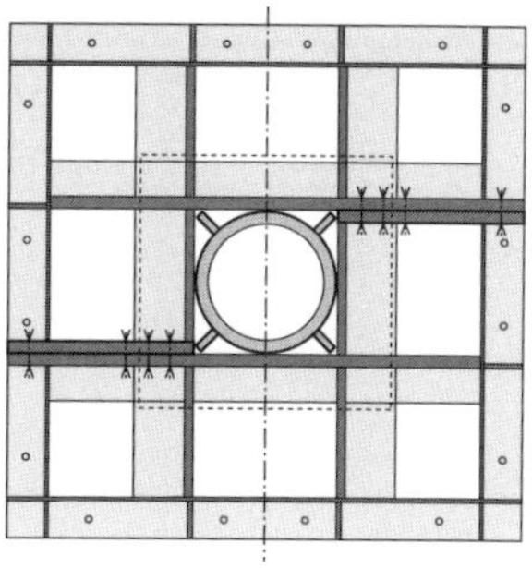

Fig. 4.3: Considered steel collar

support more dead load at less deformation. Furthermore, the load transfer into the tubular columns demanded additional measures. Apart from that, the flat slab would still fail in a brittle mode. The steel collars also would need sufficient fire protection coating. The behaviour of the steel collars at elevated temperatures becomes even softer.

4.3.1.3 Concrete collars

Two types of concrete collars (*Fig. 4.4*) were considered as alternatives to the steel collar. One with self-compacting concrete and straight tensile reinforcement with heads, and the other with shotcrete and spiral tensile reinforcement. The weight that needed to be lifted would be less than that of steel collars and more load would be carried at smaller deformations. Still, both concrete collars needed welding on site for steel supports or for additional reinforcement bars. Moreover, several construction details were not solved satisfactorily. The main arguments for the contractor were that performance and cost were unprofitable compared with steel collars.

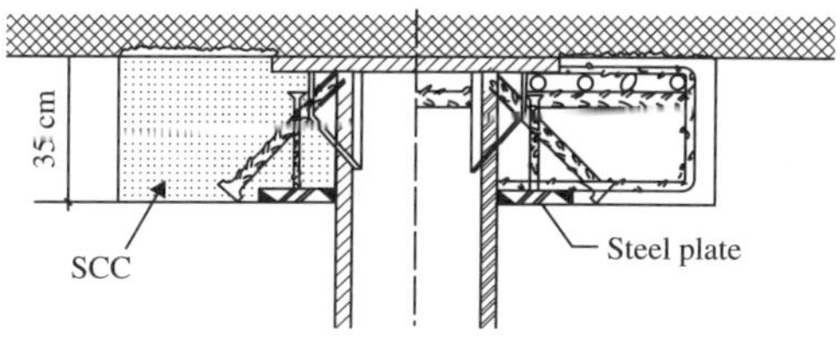

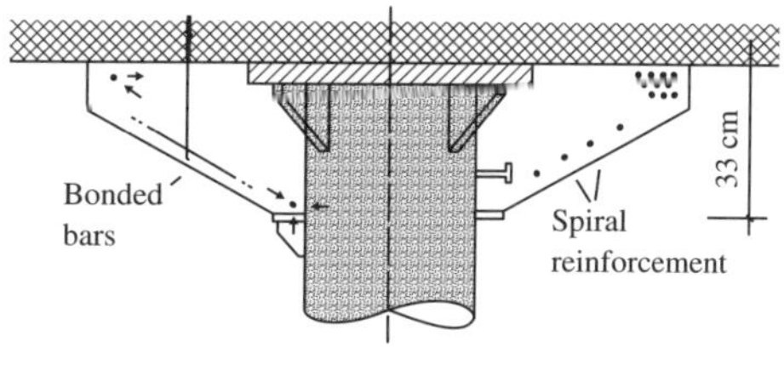

Fig. 4.4: Sketches of concrete collars with self-compacting concrete and shotcrete

4.3.1.4 Enlarged columns

If the resistance of the columns is insufficient for the given reactions, the enlargement of the columns (*Fig. 4.5*) is a suitable strengthening method. The column can be encased in two half shells of steel or be enlarged with cast-in-place reinforced concrete. Precast concrete elements are a possibility as well. Usually, the head would be widened with regard to the column. Thus, a higher punching shear resistance is achieved with the new column part supporting at least live loads. Because the columns at the Station Square in Berne have no resistance deficiency, this method was not suitable.

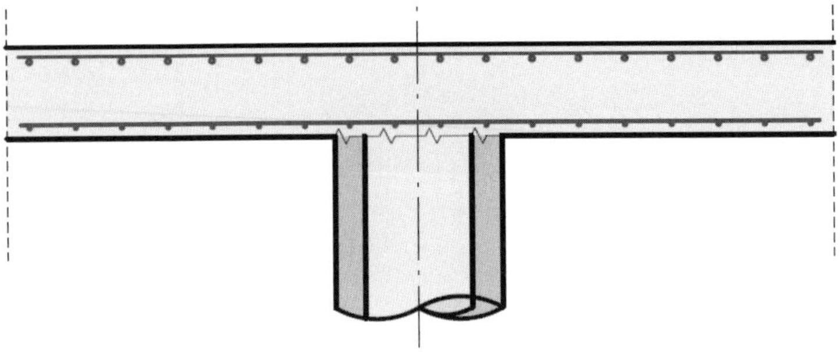

Fig. 4.5: Exemplary enlarged column [7]

4.3.1.5 Other strengthening methods

Other strengthening methods with Fibre Reinforce Polymers (FRP) or glued-on steel collars were not considered mainly because of their behaviour in case of fire and the lower strengthening potential.

4.3.2 Executed strengthening measures

4.3.2.1 Concrete overlay (executed only locally)

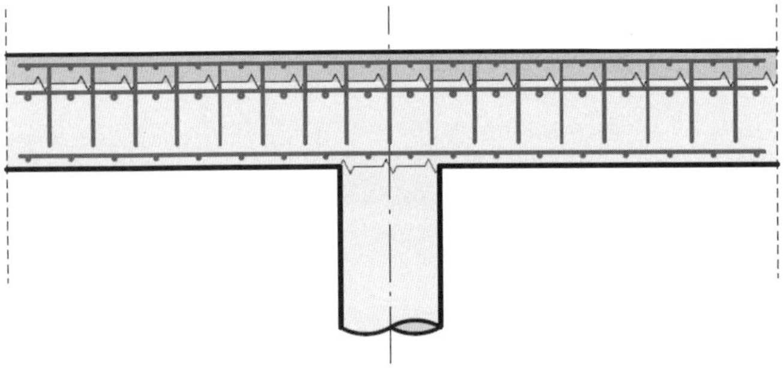

Fig. 4.6: Exemplary concrete overlay [7]

Another possibility to increase the resistance in punching shear is a concrete overlay (*Fig. 4.6*) with additional tensile reinforcement. The higher bending resistance over the column has a significant influence on punching shear resistance [5]. As mentioned before, construction time was crucial, especially on the upper side. The extensive amount of concrete, many bonded anchors to avoid delamination, and heavy and long tensile reinforcement were too many restrictions for handling in the schedule set by the contractor. Still, at some columns, this solution had to be applied because they could not be strengthened from the lower side of the slab.

4.3.2.2 Inclined post-installed punching shear reinforcement

The various restrictions on this construction site together with the disadvantages of strengthening methods discussed above demanded a new approach. Post-installed inclined punching shear reinforcement with the Hilti HZA-P anchors (*Fig. 4.7*) was chosen as the strengthening method at the New Station Square in Berne.

These anchors are used to install punching shear reinforcement into already hardened concrete. The anchors are set into hammer drilled holes under an angle of 45° towards the column after the injection of adhesive mortar Hilti HIT-RE 500. The holes should reach at least to the lower level of the upper tensile reinforcement; ideally they penetrate to its centre without damaging

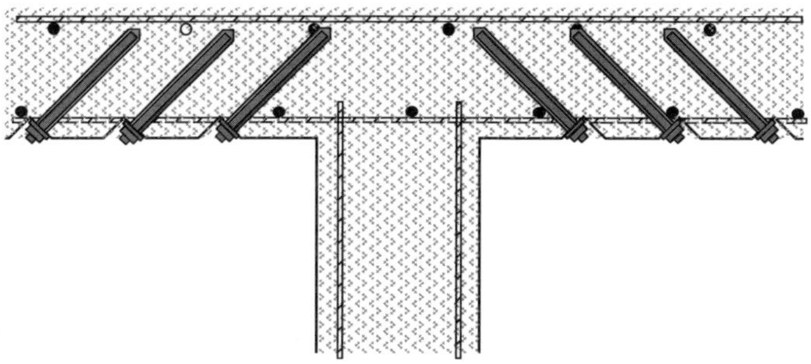

Fig. 4.7: Inclined post-installed punching shear reinforcement [8]

the reinforcement bars. The anchors consist of a reinforcement bar of diameter 20 mm and a smooth shaft with a thread. The lower part with the thread is made of higher grade steel than that of the reinforcement bar. The lower anchor head is set in the enlarged lower part of the hole after hardening of the adhesive mortar with washer and nut. The spherical washer eliminates bending of the bar. The gaps are injected through the washer with adhesive mortar. Finally, the anchor head is covered with fire protection mortar.

In this method, the construction process on the lower and upper sides of the slab is largely independent. Compared with steel or concrete collars, the method is cost efficient and the required amount of strengthening is achieved. The deformation capacity is increased considerably, and loads can thus be redistributed to neighbouring columns at ultimate limit state. Because of the embedded anchor heads, the concrete surface remains plain. Additionally, no measures have to be taken at the columns and the clearance height remains the same.

4.3.3 Durability

The durability of inclined post-installed shear reinforcement with embedded anchor heads in fire protection mortar is similar to that of cast-in-place punching shear reinforcement. Because all new elements are well embedded inside the concrete, the amount of maintenance is not increased. Existing sealing systems on the concrete slab can be left as before; new sealing systems can be applied on a plain concrete surface without interruptions. The access for chlorides to the anchors is very limited because the upper concrete surface is kept intact. The HZA-P anchors consist of carbon steel. The corrosion protection of the steel inside the adhesive has been checked and certified by the *Swiss Association for Protection against Corrosion*. The anchor head is covered by fire protection mortar also ensuring corrosion protection. Particularly for high corrosion protection, Hilti HZA-R anchors can be applied where the smooth shaft and thread are of stainless steel. The adhesive used (Hilti HIT-RE 500) has European (ETA 04/0027) and US (ICC-ES ESR 2322) technical approval that include creep and sustained load tests.

4.4 Structural analysis before and after repair, design of sections, codes

4.4.1 Structural analysis and response of structure to loads before and after repair

With the selected strengthening method, there is no significant difference regarding loads before and after repair. The modelling of the structure thus is not adjusted after repair because the system has not changed.

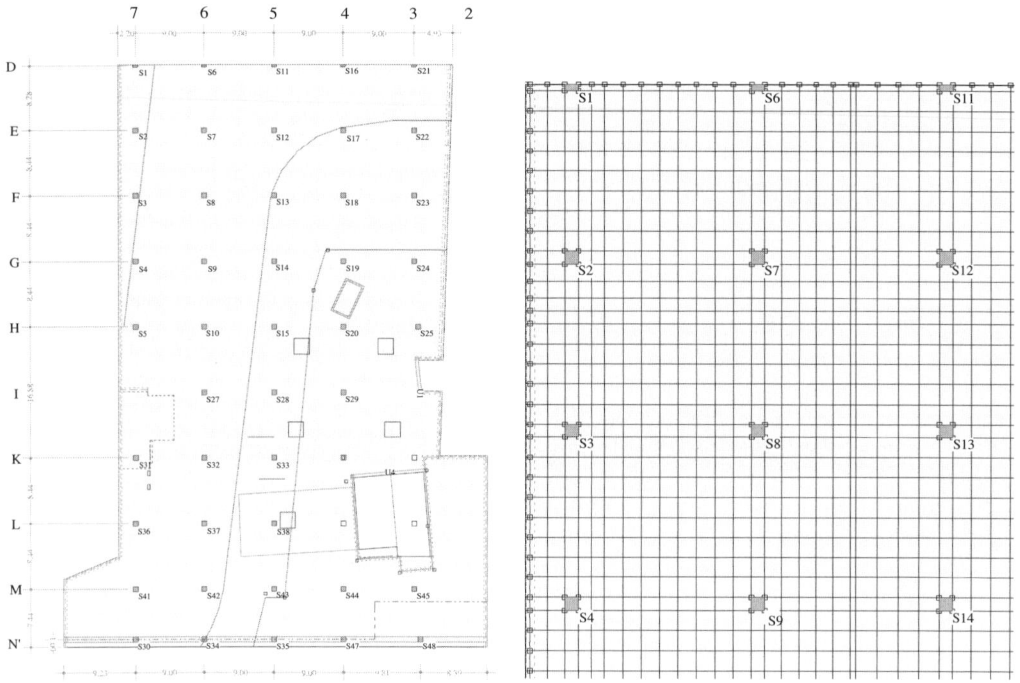

Fig. 4.8: FE model and FE grid

The first estimation of the factored shear force (V_d) is made with a simplified model with only one column and a slab with the dimensions of the regular column grid 8.44 m × 9.00 m. The obtained values can be applied for interior columns only. The order of magnitude of required strengthening thus is defined roughly based on a few calculations.

For the detailed structural analysis, the concrete slab is modelled with a finite element (FE) program (Cedrus-5, cubus engineering software, Zurich Switzerland). The slab is modelled in several elements defined by outer edges and expansion joints. *Figure 4.8* shows the central element and a section of its FE grid. The slab is supported by interior and edge columns as well as by concrete walls along the outer edge. On the edge columns along the expansion joints and on the concrete walls, the slab is supported by neoprene bearings. The slab was cast directly onto the interior columns. The vertical stiffness of concrete walls and steel columns is based on their actualized cross section, material, and height. All connections among slab, walls, and columns are modelled as hinges.

The Swiss code [4] defines a load model for road traffic with distributed loads in lanes and concentrated loads in one or two lanes depending on the road width (*Fig. 4.9*). The accompanying live load is modelled as an area and one lane with the concentrated live load as eight single loads as shown in *Fig. 4.9*. The FE model in *Fig. 4.8* is divided into a pedestrian area and a road area. The actual live load in the pedestrian area is smaller than in the road

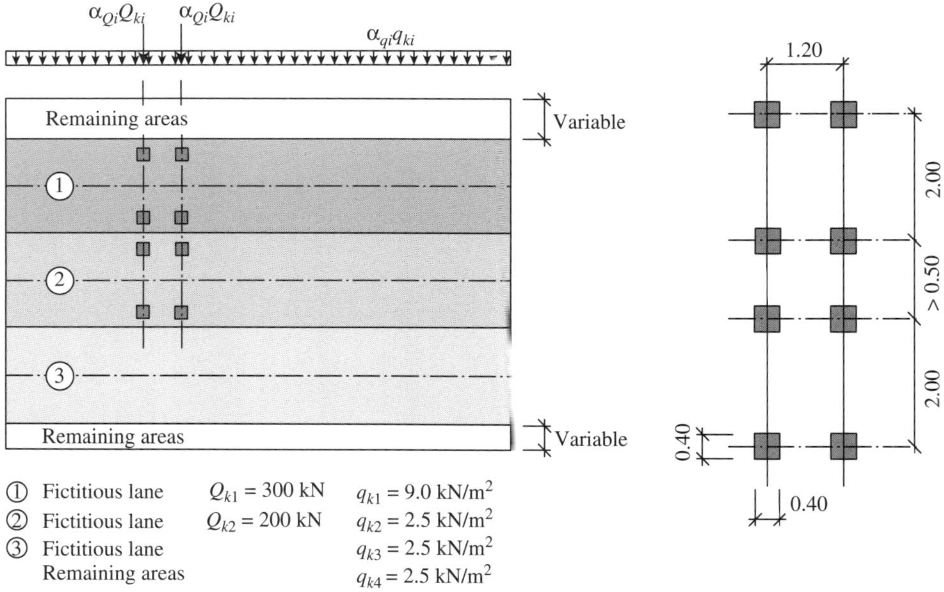

Position of load		Q_{ki} [kN]	q_{ki} [kN/m^2]
Lane 1	$(i = 1)$	300	9.0
Lane 2	$(i = 2)$	200	2.5
Lane 3	$(i = 3)$	–	2.5
Other lanes	$(i > 3)$	–	2.5
Remaining areas	$(i = r)$	–	2.5

Fig. 4.9: Applied load model, possible arrangement, and characteristic values [4]

Actions	γ_F	Limit state		
		Type 1	Type 2	Type 3
Permanent actions				
- Acting unfavourably	$\gamma_{G,sup}$	1.10 1)	1.35 1)	1.00
- Acting favourably	$\gamma_{G,inf}$	0.90 1)	0.80 1)	1.00
Variable actions				
- In general	γ_Q	1.50	1.50	1.30
- Road traffic loads	γ_Q	1.50	1.50	1.30
- Rail traffic loads	γ_Q	1.45	1.45	1.25
Actions imposed by the ground				
Soil surcharge				
- Acting unfavourably	$\gamma_{G,sup}$	1.10	1.35 2)3)	1.00
- Acting favourably	$\gamma_{G,inf}$	0.90	0.80	1.00
Earth pressure				
- Acting unfavourably	$\gamma_{G,Q,sup}$	1.35	1.35	1.00
- Acting favourably 4)	$\gamma_{G,Q,inf}$	0.80	0.70	1.00
Water pressure				
- Acting unfavourably	$\gamma_{G,Q,sup}$	1.05	1.20 3)	1.00
- Acting favourably	$\gamma_{G,Q,inf}$	0.95	0.90	1.00

1) G is either multiplied by $\gamma_{G,sup}$ or by $\gamma_{G,inf,}$ depending on whether the overall action effect is unfavourable or favourable.

2) For heights of fill from 2 to 6 m, $\gamma_{G,sup}$ may be reduced linearly from 1.35 to 1.20.

3) When using the observational method, according to SIA 267, reduced values are admissible in certain cases.

4) For passive earth pressure acting favourably, $F_d = R_d$ according to SIA 267.

Fig. 4.10: Load factors γ_F for the verification of structural safety according to Ref. [3]

area and thus accounted for by a lower factor according to Ref. [4]. Because of the gravel road bed, dynamic impact on the concrete slab is limited. Accordingly, the live loads are accounted for by a reduced dynamic factor.

The partial load factors are applied according to Ref. [3] (*Fig. 4.10*). The resulting factored column reactions at design level are shown in *Fig. 4.11*. Because of the different and varying loads (road area and pedestrian area, road bed thickness, etc.), the values of the resulting force vary considerably. To obtain the predominant factored shear force V_d, the dead load within the control perimeter is subtracted.

The reconstructed Station Square is also reorganized. New staircases are needed and old ones have to be closed. Additionally, a few columns are eliminated and expansion joints in the concrete slab are closed for durability reasons. Besides the strengthening against punching shear, the concrete slab is strengthened, expanded, cut, and mounted on new bearings in various ways. However, the description of all concrete works carried out at the New Station Square in Berne is beyond the scope of this paper.

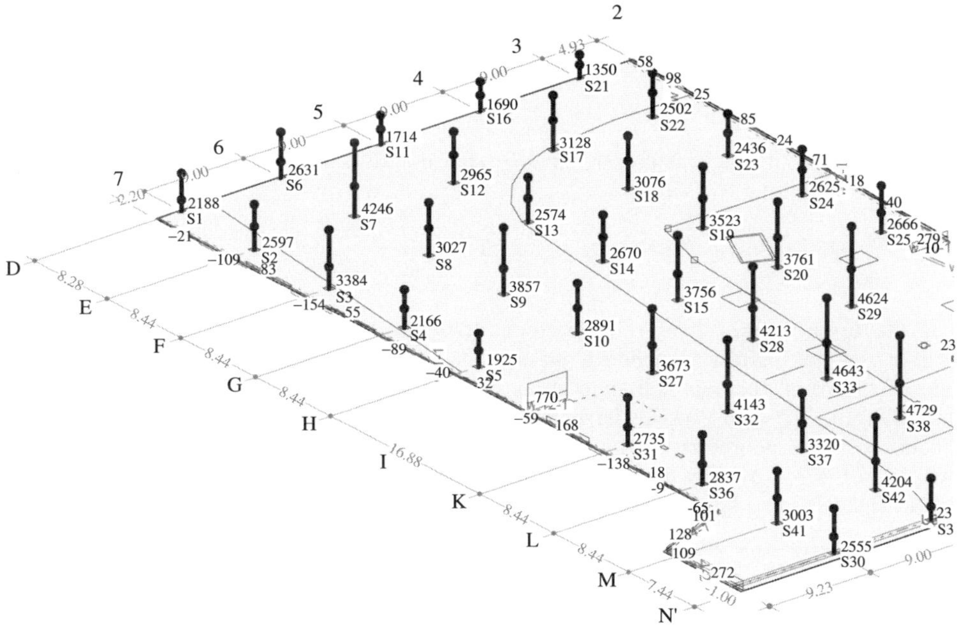

Fig. 4.11: Factored column reactions

4.4.2 Design of sections (section analysis) before and after repair

4.4.2.1 Before repair

The concrete slab contains no punching shear reinforcement. The insufficient punching shear strength before repair is already mentioned in the tender documents. Still, for the assessment of the structural safety during construction, calculations regarding the condition before repair were conducted. On the basis of the failure criterion and load–rotation relationship as mentioned in Ref. [9], the actual deficiency in punching shear resistance can be reduced if calculated iteratively. On the other hand, the live and dead loads can be restricted on the construction site to guarantee structural safety by a sufficient margin during construction.

4.4.2.2 After repair

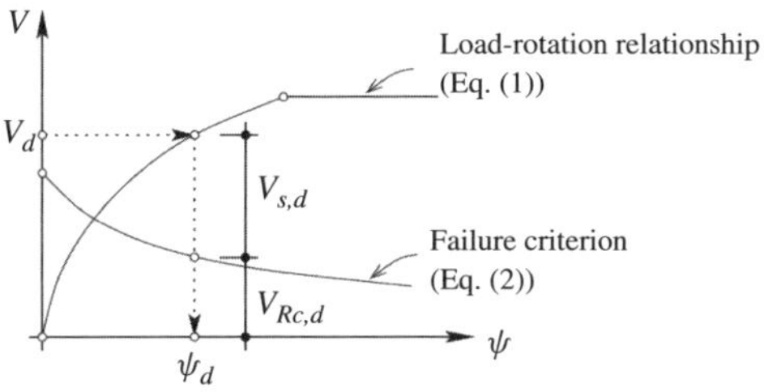

Fig. 4.12: Load–rotation relationship and failure criterion [9]

The design method for the post-installed punching shear reinforcement is based on the critical shear crack theory as mentioned in Refs. [9, 10]. According to this theory, the rotation of the slab will eventually open a shear crack inside. The rotation can be expressed as function of the column load. *Figure 4.12* shows the load–rotation relationship and the failure criterion. The contribution of concrete at failure can be estimated according to Ref. [9]. The remaining design load is assumed to be carried by the punching shear reinforcement.

According to Ref. [9], the shear reinforcement is designed to satisfy the following condition.

$$V_{s,\mathrm{d}} \leq \sum_{i=1}^{n} N_{si,\mathrm{d}} \times \sin \beta_i \tag{4.1}$$

The factored strength of one tension anchor ($N_{si,d}$) is equal to the minimum of the following values:

$$N_{si,\mathrm{d}} = \min(N_{si,el,\mathrm{d}}; N_{si,pl,\mathrm{d}}; N_{si,b,\mathrm{d}}; N_{si,p,\mathrm{d}}) \tag{4.2}$$

$N_{si,el,d}$ is the force in the shear reinforcement that can be activated assuming an elastic behaviour in the bar mainly depending on the rotation of the slab. $N_{si,pl,d}$ is the plastic resistance of the reinforcement bar, and $N_{si,b,d}$ is the upper limit of the resistance due to the bond strength. $N_{si,p,d}$ is the resistance against pullout by concrete cone failure of the lower anchorage. The geometry of the reinforcement and the basic variables for calculation are shown in *Fig. 4.13*; usually the anchors are installed at an angle $\beta = 45^\circ$. Further details are given in Refs. [8, 9, 11].

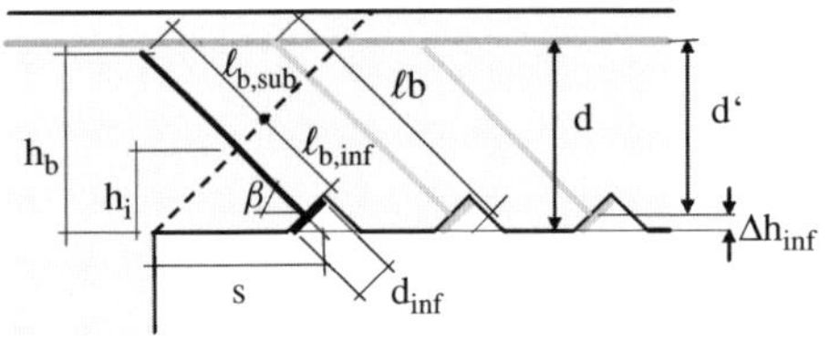

Fig. 4.13: Geometry of reinforcement [11]

Even with punching shear reinforcement, the codes usually define a maximum punching shear strength ($V_{\mathrm{Rd,max,code}}$) accounting for failure of the compression zone of the slab near the column. The design method for the post-installed punching shear reinforcement with Hilti HZA-P anchors also defines a maximum resistance that can be achieved [9]. This value should not be exceeded even if $V_{\mathrm{Rd,max,code}}$ is higher.

The reinforced area must be such that the verification for the outer control perimeter guarantees sufficient punching shear resistance outside the reinforced area according to the applied structural concrete code. It should be mentioned that the static height is reduced if the lower anchor head is embedded in the concrete slab. *Figure 4.14* shows the definition of the outer control perimeter. It can be increased with intermediate anchors in the outermost circle.

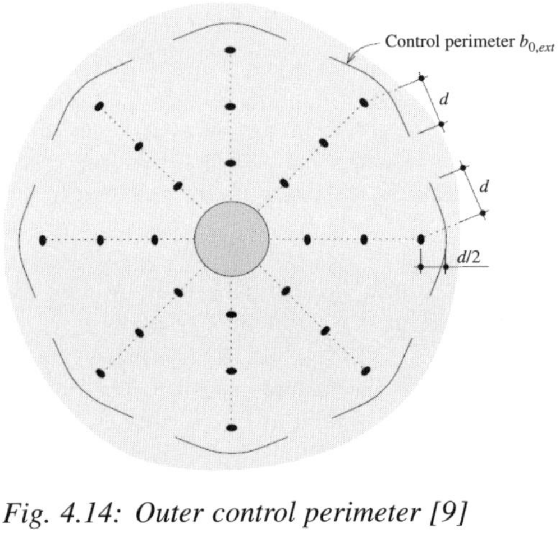

Fig. 4.14: Outer control perimeter [9]

Hilti developed the EXBAR punching design software for the strengthening of structural parts against punching shear. It is based on the design method presented in Ref. [9]. For the application at the New Station Square in Berne, the software was not yet available. Thus, an internal version had to be developed for this first application. The later comparison with EXBAR showed no differences in the calculated punching shear resistance. The inclined post-installed punching shear reinforcement allows for an increase in punching shear resistance up to 60%. The average requested strengthening ratio is around 30% with a few exceptions.

Because all elements of the strengthening method are inside the structure, no special calculations or measures apart from fire protection mortar have been undertaken for the resistance in the case of fire. The anchors including the lower head are treated as the cast-in-place tensile reinforcement. Fatigue of the strengthening method is a minor issue. The live load is only 10–30% of the total dead load. Thus, the amount of stress reversal is limited. Nevertheless, the design of bonded-in reinforcement bars for predominantly cyclic (fatigue) loading with adhesive mortar is shown in Ref. [12]. The tests with hammer drilled holes and purely pulsating load mentioned in Section 3.7 show no fatigue failure of the bond. The design value of the bond strength with adhesive mortar HIT-RE 500 has to be calculated according to Ref. [11] taking the effective concrete strength into account.

The detailing rules are shown in *Fig. 4.15*. The post-installed shear reinforcement anchors have to be set around the column in at least two circles. The angle α_h between the radii where the anchors are set may be maximally 45°. The radial distance between two anchors s_1 and between the first anchor and the edge of the supported area s_0 should be less than $0.75d$. The value of s_0 should not be selected too small, otherwise the capacity of the first reinforcement bar may be reduced or execution difficulties may occur because of dense column reinforcement. The direction of the hammer drilled holes should be at an angle $\beta_i = 45°$ and towards the column. The anchor should be bonded up to the height h_b, ideally equal to the average static height d of the tensile reinforcement.

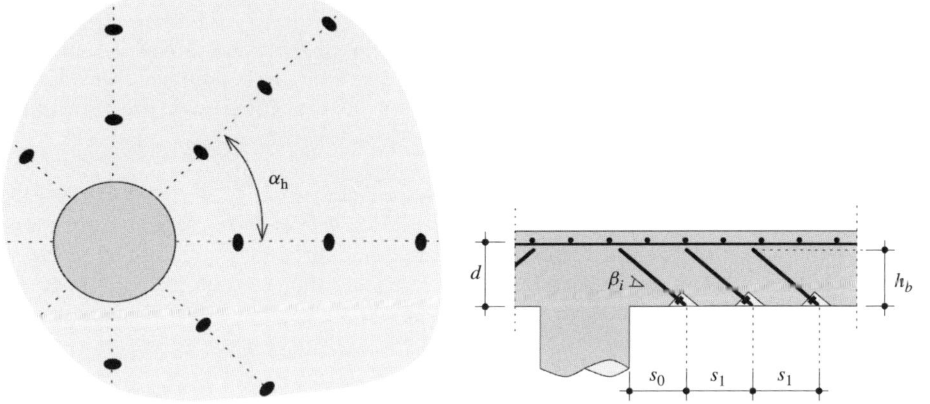

Fig. 4.15: Detailing rules of the design method [9]

4.4.3 Codes

The Swiss codes SIA 260–267 introduced in 2003 systematically comply with the principle of verification on design level (*Fig. 4.16*). The same applies to the Eurocode, although the reduction and load coefficients are not exactly the same. Generally, the Eurocode covers more design situations, but the Swiss codes provide enough information for the design of most structures. Because the Swiss code is based on the Eurocode, it is possible to combine the two codes. Nevertheless, design based on one code is advisable.

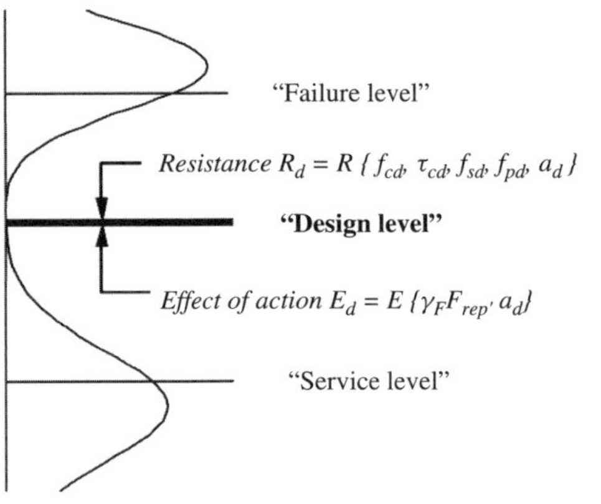

Fig. 4.16: Verification on design level (translated from Ref. [13])

The system of post-installed shear reinforcement does not comply with two articles in the Swiss code for structural concrete [5] concerning bond properties. According to Ref. [3], exceptions are permissible if they are based on established, if necessary experimentally confirmed, theory and engineering practice. The essential critical shear crack theory for this strengthening method is described in Refs. [9, 10]. Additionally, Hilti conducted experimental tests on slabs as described in Section 3.7 and Ref. [11].

4.5 Detailing

The strengthening anchor consists of a reinforcement bar of diameter 20 mm in the upper part. The lower part is a smooth shaft with a thread at the end (*Fig. 4.17*). For the design, the strength of the reinforcement bar is decisive, because the smooth shaft and thread are made of a high grade steel. The lower head is a spherical washer to eliminate bending of bar and nut (*Fig. 4.18*). The enlarged part of the hole with the anchor head is filled with fire protection mortar.

For strengthening against punching shear at the New Station Square in Berne, the anchors are post-installed around the column in two or three circles *(Fig. 4.19).* Only these two different arrangements depending on the required amount of strengthening were defined. The distance from the column centre point to the first circle is 620 mm, and the distance to the next circle is 300 mm (approximately 0.5d). The angle between two radii is 30°. 24 or 36 anchors are applied per column, leading to a total of 1836 anchors.

The lower layer of the upper reinforcement is only 4.8 m long. According to Ref. [5], a length of 3d not including development length beyond the supported area is necessary. With the post-installed punching shear reinforcement, this specification of reinforcement length beyond supported area is not complied with (*Fig. 4.20*). To avoid crack opening along the upper tensile reinforcement in the case of punching and extraction of the upper tensile reinforcement layer, two vertical anchors are set from above towards the end of the reinforcement bars).

Fig. 4.17: Elements of the Hilti strengthening anchor HZA-P [11]

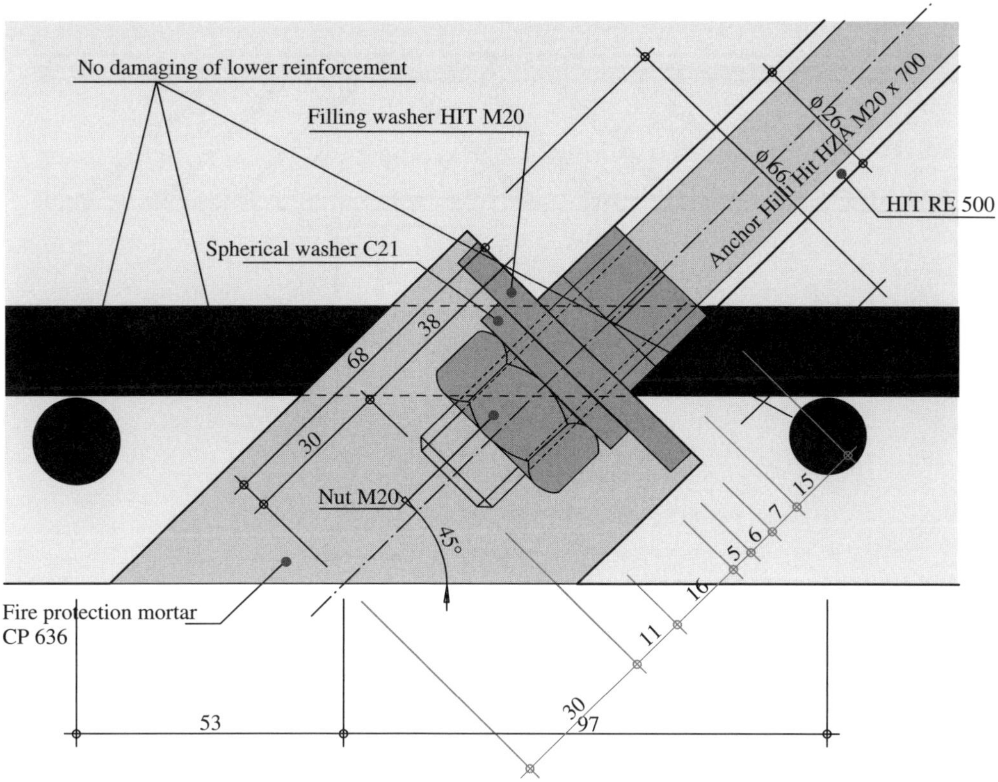

Fig. 4.18: Construction detail, lower anchor head (Units: mm)

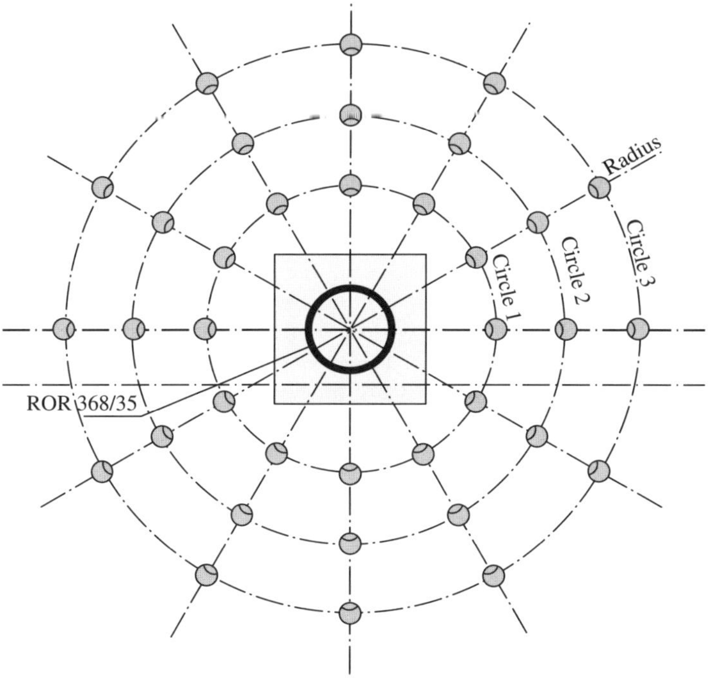

Fig. 4.19: Arrangement of anchors in plan

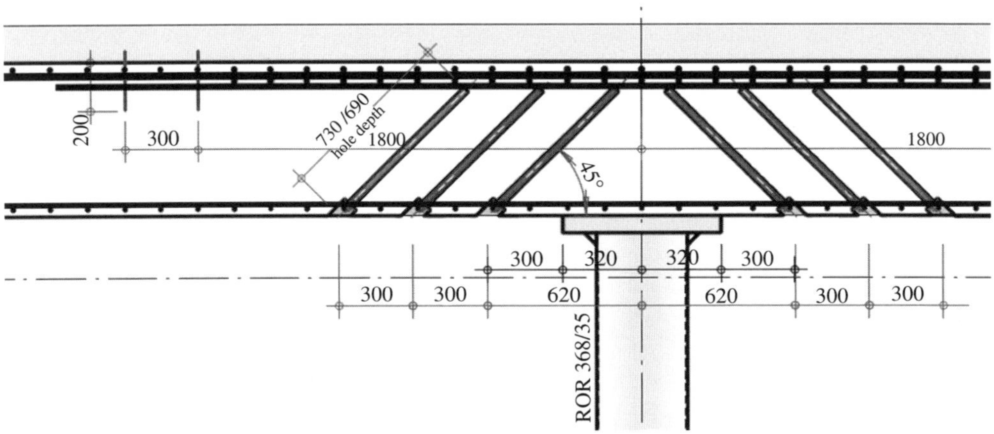

Fig. 4.20: Cross section of the strengthening (Units: mm)

4.6 Construction procedure

The construction process and the equipment are illustrated in *Figs. 4.21–4.25*. The selected strengthening method allows independent construction on the upper and lower sides of the concrete slab. Only the anchoring of the upper tensile reinforcement layer has to be executed from the upper side. The power drill is either mounted on a one-footed trestle or anchored to the slab.

Fig. 4.21: Mobile scaffold and hammer drill anchored to the slab

The inclined holes are hammer drilled from the bottom into the concrete slab under an angle of 45° towards the column. The lower end of the hole has to be enlarged with a special drill (*Fig. 4.23*) for the later installed anchor head. Thus, the lower tensile reinforcement has to be marked previously to avoid damage. The drilled holes ideally should end between the two layers of upper tensile reinforcement in the two directions, and it should reach at least the lower level of the tensile reinforcement.

The hammer drilled holes are injected with adhesive mortar. The described strengthening anchors are set into the holes subsequently. The lower anchor head is installed after hardening of the adhesive mortar. To ensure a slip-free anchorage, the interface between washer and concrete is injected with adhesive mortar through the washer. Finally, the enlarged lower part of the hole is filled with fire protection mortar. The anchor head is thus completely embedded and not visible.

Fig. 4.22: Drilling for anchoring of upper tensile reinforcement and for punching shear reinforcement

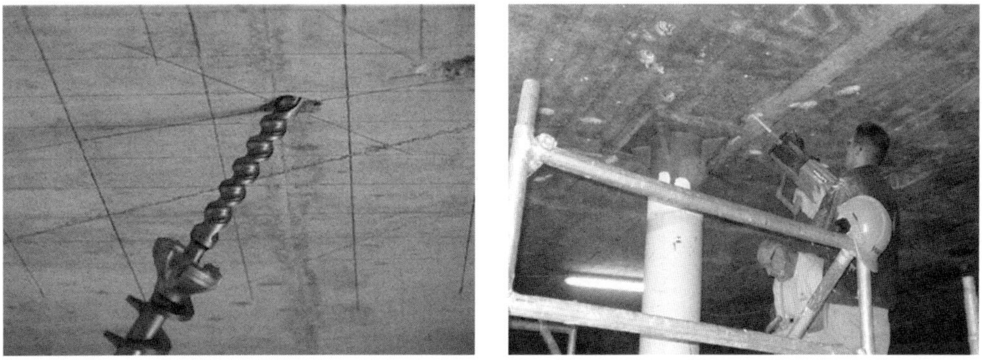

Fig. 4.23: Enlargement of the hole for anchor heads and filling the holes with adhesive mortar

Fig. 4.24: Adjusted HZA-P anchors to the individual hole depth and installing in the hole

Because of the decreasing resistance with holes not reaching the required length, it is important to ensure continuous construction inspection. The actual depth of the hole should be noted next to it on the slab. The anchors have to be shortened to the respective length of the hole they are installed in. If it is not possible to reach the requested minimum depth, the unfinished hole has to be filled with adhesive mortar and a short piece of reinforcement bar. For each

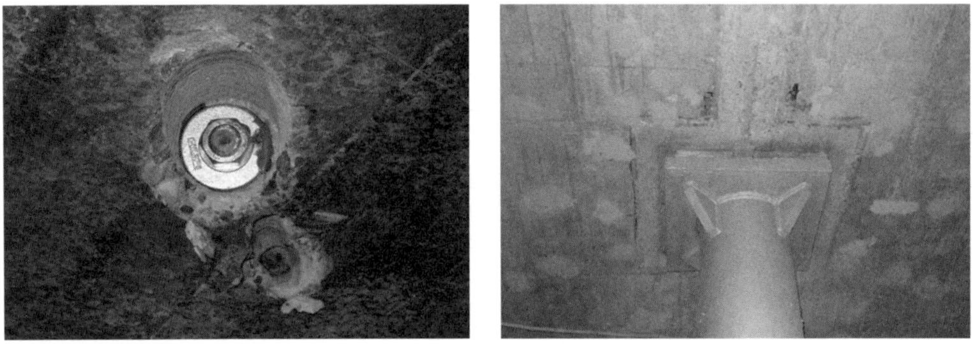

Fig. 4.25: Embedded and injected anchor head next to an unfinished hole and concrete surface after strengthening

circle of post-installed anchors, the dependency of achievable resistance and actual hole depth is different. Thus, a minimum and an allowable margin for the hole depth has to be defined for each circle. The outermost circle has the smallest margin. Another tolerance to be defined is the lateral deviation. Because unfinished holes cannot be avoided, the final anchor will not be exactly at the predefined location (*Fig. 4.25*). The execution of the strengthening is not possible without reasonable tolerances. Nevertheless, the safety against punching shear failure has to be assured. The construction inspection has to collect all hole depths, anchor lengths, lateral deviations, and the amount of unfinished holes. Values exceeding the given tolerance have to be investigated by the engineer.

At the New Station Square in Berne, 12 holes had to be drilled around the columns in two or three circles, which amount to a total of 24 or 36 holes for the columns, with regard to the required strengthening. These holes result in a considerable weakening during the construction process. For a maximized strengthening in the final state and for safety reasons during the construction process, it is necessary to minimize the load on the column during construction. Additionally, the drilling was divided into two stages. Every second hole in a circle was left out until the first half of the anchors were installed and the adhesive mortar hardened. With these two stages, no temporary shoring was necessary.

4.7 Load testing

It was not possible to conduct load testing on site. Experiments confirming the efficiency of strengthening concrete slabs against punching shear with post-installed shear reinforcement were conducted at the Hilti Laboratory in Schaan and evaluated together with Professor Dr A. Muttoni from the Swiss Federal Institute of Technology in Lausanne, Switzerland. Slabs of 3 m × 3 m × 0.25 m were subjected to monotonically increasing punching shear load. Various ratios of tensile and shear reinforcement were investigated. *Figure 4.26* shows the test setup.

The load was applied from bottom to top by a hydraulic cylinder. Deformation of the slab, strain on the concrete, and strain in tensile and shear reinforcement were measured. *Figure 4.27* shows a sample load–displacement curve. The increased resistance for punching shear and the higher deformation capacity are clearly visible. It shows the results for a slab that is not reinforced, a slab with post-installed inclined punching shear reinforcement (Hilti test

Fig. 4.26: Test setup [8]

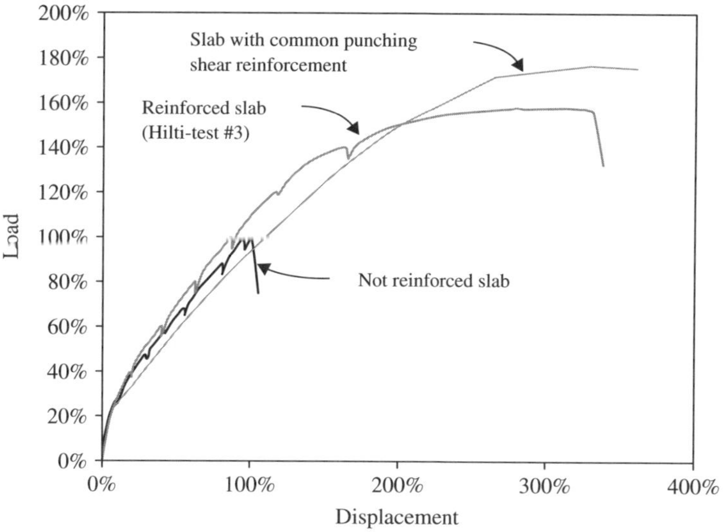

Fig. 4.27: Load displacement curves (translated from Ref. [7])

no. 3), and a slab with common cast-in-place punching shear reinforcement. *Figure 4.28* shows the failure patterns for concrete slabs with and without post-installed punching shear reinforcement. The Hilti internal test reports will not be published, for more details see Refs. [8, 11].

Fig. 4.28: Failure patterns with and without post-installed punching shear reinforcement [11]

4.8 Summary

Over the years, several needs developed in the surroundings of the old Station Square in Berne. The overlapping of public and individual transportation needed to be reorganized including its structures. The underground passage had structural safety and functional deficiencies at several elements. Because of the interdependencies, an overall solution had to be developed. The insufficient punching shear strength was identified during the preparation of the tender documents. More stringent code requirements and additional loads are the main causes for the deficiency.

As structural engineer for the contractor, the authors considered conventional strengthening methods at first. Vertical post-installed punching shear reinforcement with anchor heads on both sides of the slab is an efficient and reliable method. However, the dependency of construction works on the upper and lower sides restrict the construction process to a great extent. Steel and concrete collars can be mounted at the column heads; however, construction performance, limited deformation capacity, and costs are unfavourable. The strengthening with a concrete overlay, anchoring, and additional tensile reinforcement demands extensive work on the upper side. Because of the short time available, this method was only applied locally. Finally, the Swiss Federal Institute of Technology in Lausanne and Hilti made the first application of a newly developed strengthening method possible. The Hilti HZA-P strengthening anchors work as post-installed inclined punching shear reinforcement. They are set into hammer drilled holes around the column with adhesive mortar. The inclined bonded bars increase the punching shear resistance and the deformation capacity considerably. Thus, the safety of the whole structure is enhanced because loads can be redistributed to neighbouring columns. Additionally, the clearance in the underground passage is not reduced. A permanent construction inspection is essential to guarantee the designed punching shear resistance. With the application at the New Station Square in Berne, the method proved to be suitable and cost efficient.

4.9 Acknowledgements

The first application of the post-installed inclined punching shear reinforcement with Hilti strengthening anchors HZA-P as reported herein was only possible with the approval of the client (Public Work Service, City of Berne) and the straightforward collaboration with

Prof. Dr A. Muttoni (Ecole Polytechnique Fédérale de Lausanne, Switzerland) and Dr J. Kunz (Hilti Corp., Liechtenstein). The authors thank Prof. Dr A. Muttoni for calling their attention to this new strengthening method and his expertise during design and execution. They also thank Dr J. Kunz and Hilti for their contribution to the project and Prof. Dr D. Zwicky for supporting this paper.

References

[1] SIA 160:1970. *Norm 160: Norm für die Belastungsannahmen, die Inbetriebnahme und die Überwachung der Bauten*. Schweizerischer Ingenieur- und Architektenverein (SIA), Zürich, 1970.

[2] SIA 162:1968. *Norm 162: Norm für die Berechnung, Konstruktion und Ausführung von Bauwerken aus Beton, Stahlbeton und Spannbeton*. Schweizerischer Ingenieur- und Architektenverein (SIA), Zürich, 1968.

[3] SIA 260:2003 (e). *Code 260: Basis of Structural Design*. Schweizerischer Ingenieur- und Architektenverein (SIA), Zürich, 2003.

[4] SIA 261:2003 (e). *Code 261: Actions on Structures*. Schweizerischer Ingenieur- und Architektenverein (SIA), Zürich, 2003.

[5] SIA 262:2003 (e). *Code 262: Concrete Structures*. Schweizerischer Ingenieur- und Architektenverein (SIA), Zürich, 2003.

[6] prEN 13791:2006 D. *Bewertung der Druckfestigkeit von Beton in Bauwerken oder in Bauwerksteilen (Assessment of In-situ Compressive Strength in Structures and Precast Concrete Components)*. Europäisches Komitee für Normung (CEN), Brüssel, Juni 2006 (final draft version).

[7] SIA. *Dokumentation D0226: Tragsicherheit von Einstellhallen*. Schweizerischer Ingenieur- und Architektenverein (SIA), Zürich, 2008.

[8] Kunz J, Fernandez Ruiz M, Muttoni A. Enhanced safety with post-installed punching shear reinforcement. *fib International Symposium*, Amsterdam, The Netherlands, 2008.

[9] Muttoni A, Fernandez Ruiz M. *Design Method for Post-Installed Punching Shear Reinforcement with Hilti Tension Anchors HZA*. Ecole Polytechnique Fédérale, Lausanne, 2007.

[10] Muttoni A, Fernandez Ruiz M. Shear strength in one- and two-way slabs according to the critical shear crack theory. *fib International Symposium*, Amsterdam, The Netherlands, 2008.

[11] HILTI. *Hilti Fastening Technology Manual - Post-Installed Shear Reinforcement Hilti HZA-P*. HILTI Corporation, Schaan, 2009.

[12] HILTI. *Design of Bonded-in Reinforcement (Post-installed Rebar Connections) Using Hilti HIT-HY 150 or Hilti HIT-RE 500 for Predominantly Cyclic (fatigue) Loading*. HILTI, Schaan, 2002.

[13] SIA. *Dokumentation D0182: Einführung in die Norm SIA 262*. Schweizerischer Ingenieur- und Architektenverein (SIA), Zürich, 2003.

Chapter

5

Strengthening of the Frame Structure at the Timisoreana Brewery, Romania

Corneliu Bob, Professor; **Sorin Dan**, Lecturer, Dr; **Catalin Badea,** Lecturer, Dr, Department of Civil Engineering, "Politehnica" University of Timisoara, Timisoara, Romania; **Aurelian Gruin,** Researcher, Eng., Building Research Institute INCERC, Timisoara, Romania and **Liana Iures**, Assistant Professor, Dr, Department of Civil Engineering, "Politehnica" University of Timisoara, Timisoara, Romania

Abstract: Many structures built in Romania before 1970 were designed for gravity loads with inadequate lateral load resistance because earlier codes specified lower levels of seismic loads. Some of these structures are still in service beyond their design life. Also, some deterioration was observed in existing structures due to the actions of different hazard factors. This paper presents the case study of a brewery with reinforced concrete framed structure of five storeys and a tower of nine storeys, which has been assessed and strengthened. The brewery and the tower were built in 1961 and an extension in 1971. An assessment performed in 1999 showed up local damages at slabs, main girders, secondary beams, and columns; concrete carbonation; concrete cover spalled over a large surface; complete corrosion of many stirrups and deep corrosion of main reinforcement; and some broken reinforcement. Such damage was caused by salt solution, CO_2, relative humidity RH $\approx$ 80%, and temperatures over 40°C. Also, inadequate longitudinal reinforcement was deduced from the structural analysis. The initial design, done in 1960, was according to the Romanian codes of that time with provisions at low seismic actions. The structural system weakness is due to present-day high seismic actions. The rehabilitation of the reinforced concrete structure was performed by jacketing with reinforced concrete for the main and secondary beams and columns. In 2003, due to continuous operation and subsequent damage of the structure, a new assessment was required. It was found that some beams and one column were characterized by inadequate main and shear reinforcement as well as corrosion of many stirrups at beams. The strengthening solution adopted was based on carbon fibre reinforced polymer composites for beams and column.

Keywords: existing reinforced concrete structures; reinforcement corrosion; seismic action; assessment and rehabilitation; structural analysis; strengthening solutions.

5.1 Introduction

Assessment of the protection level of structures, generally and particularly of reinforced concrete structures, has become a constant preoccupation of many specialists involved in design, execution, and monitoring of structures. For achieving this goal, it is necessary to estimate quantitatively two parameters: durability and safety, which are the principal components of construction quality.

The structural durability may be defined as the time period during which the construction preserves its own normal characteristics of function. The structural safety has to take into account the effect of all possible actions, ordinary loads, and extreme loads: permanent, variable and extreme actions and the environmental factors.

Different factors, which affected the existing structure presented in this paper, are reinforcement corrosion due to carbonation and chloride penetration and seismic action.

Reinforced concrete structures that are subjected to environmental conditions are likely, after a certain period of exposure, to exhibit signs of distress as a result of initiation of reinforcement corrosion process. The initial corrosion occurs mainly in two different ways: carbonation of the concrete surrounding the reinforcement and presence of chloride. The principal correlation, which characterizes the reinforcement corrosion—an important part of concrete durability—is the depth of carbonation or chloride penetration and the time of CO_2 or/and Cl^- action. Main factors influencing carbonation and chloride ingress are carbonation dioxide and chloride concentration, environmental conditions, permeability properties, and chemical reaction.

The vulnerability of existing structures under seismic motions may be due to structural system weaknesses and specific detailing [1–4]. Structural weaknesses are characterized by various irregularities and discontinuities or by general structural vulnerabilities:

a. Irregularities in the vertical direction of the buildings: irregular distributions of the stiffness; strength discontinuities; mass irregularities; vertical load discontinuities.
b. Irregularities in the building layout: horizontal irregularities of mass, stiffness, and strength, all of which produce torsion effects; unfavourable plan layouts; slab discontinuities due to holes or weaknesses of the connections in some zones.
c. General structural vulnerabilities: the indirect transfer of strong forces by beam-on-beam supports or columns supported on beams; cantilever horizontal members with large spans and/or high loads; weak column/manageable strong beam; eccentricities; finite service life due to deterioration of the constituent parts of the building.

Reinforced concrete structures may be characterized by common non-ductile detailing and vulnerabilities [1–3]:

- inadequate column bending and shear capacity;
- inadequate beam shear resistance;
- inadequate joint shear resistance;

Fig. 5.1: The Timisoreana Brewery—main facade

- inadequate quantities and anchorage of beam-positive reinforcement at the beam–column joint;
- inadequate confinement of the potentially plastic hinges of the columns and beams as well as of the boundary elements of reinforced concrete frame-wall systems;
- inadequate reinforcement of the frame in the longitudinal direction of the building.

5.1.1 Description of structure

The Timisoreana Brewery, a reinforced concrete framed structure with one section of five storeys and a tower of nine storeys, *Fig. 5.1*, has been assessed and strengthened in two steps, in 1999 and in 2003. The brewery and the tower were built in 1961, and the extension in 1971.

The industrial building vertical structure is a spatial frame as detailed in *Figs. 5.2 and 5.3*. The foundation system consists of isolated reinforced concrete foundations under columns.

The reinforced concrete monolithic floors are made of secondary and main beams and a one way reinforced slab, as shown in *Fig. 5.4*.

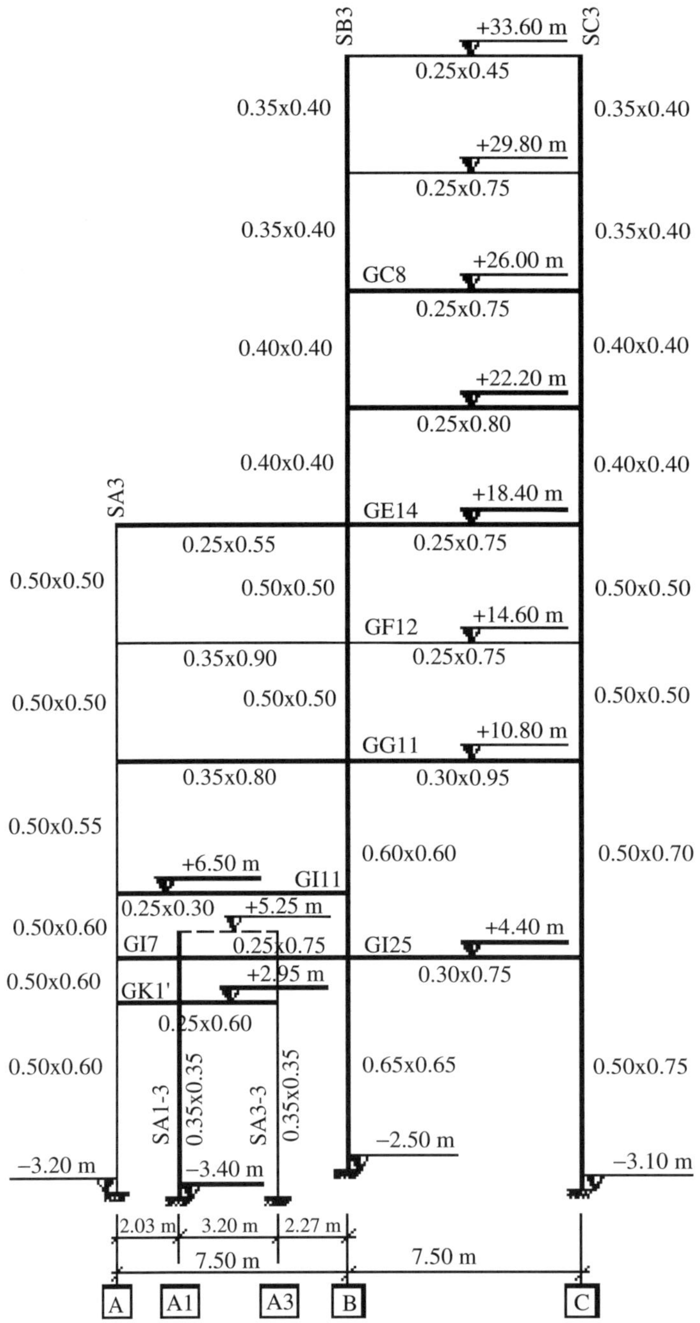

Fig. 5.2: Transversal frame (Units: m)

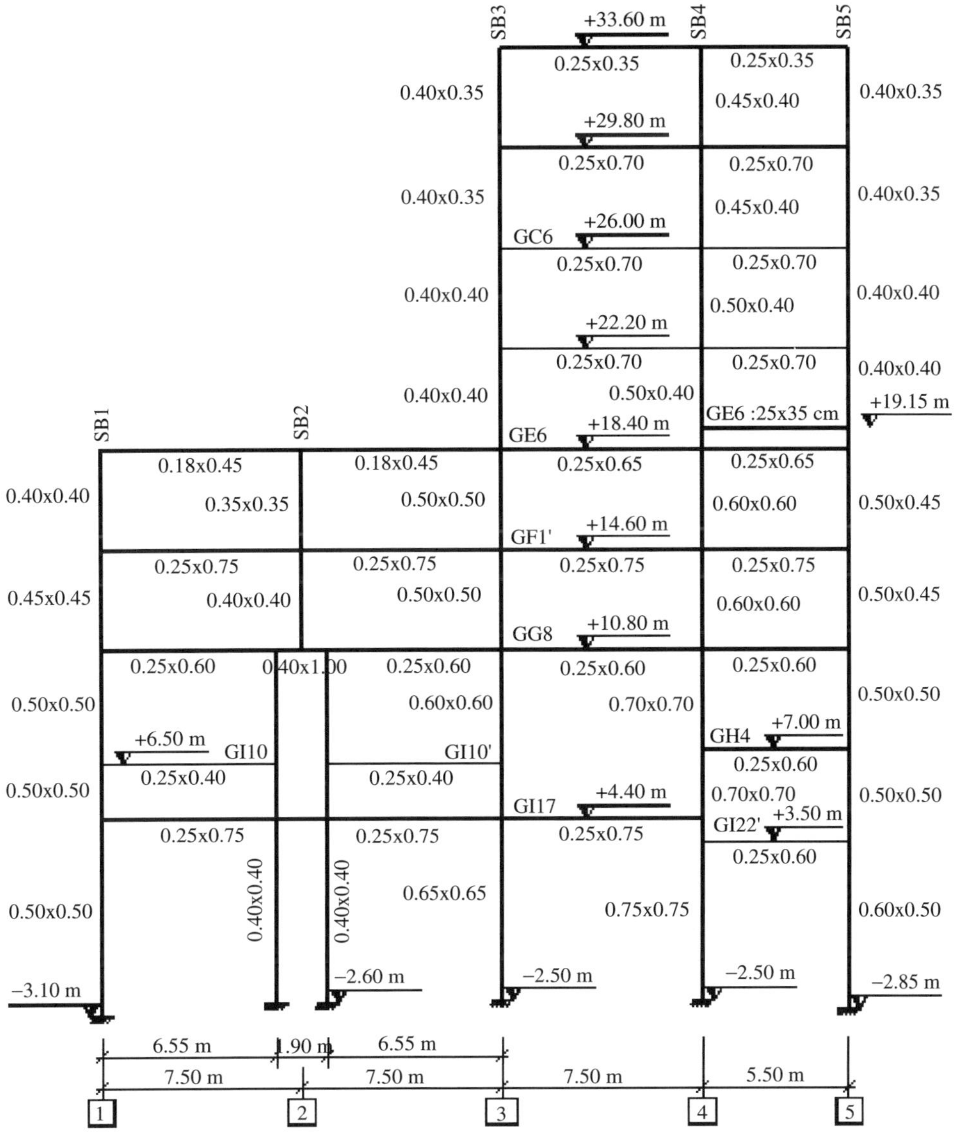

Fig. 5.3: Longitudinal frame (Units: m)

5.2 Symptoms that led to need of strengthening and assessment of *in situ* conditions

5.2.1 Symptoms that led to need of strengthening

The main problems were local damage of some structural elements and inadequate reinforcement of columns and beams at seismic actions. Local damage was noticed and assessed at slabs, main girders, secondary beams, and columns. The damage consisted of concrete carbonation:

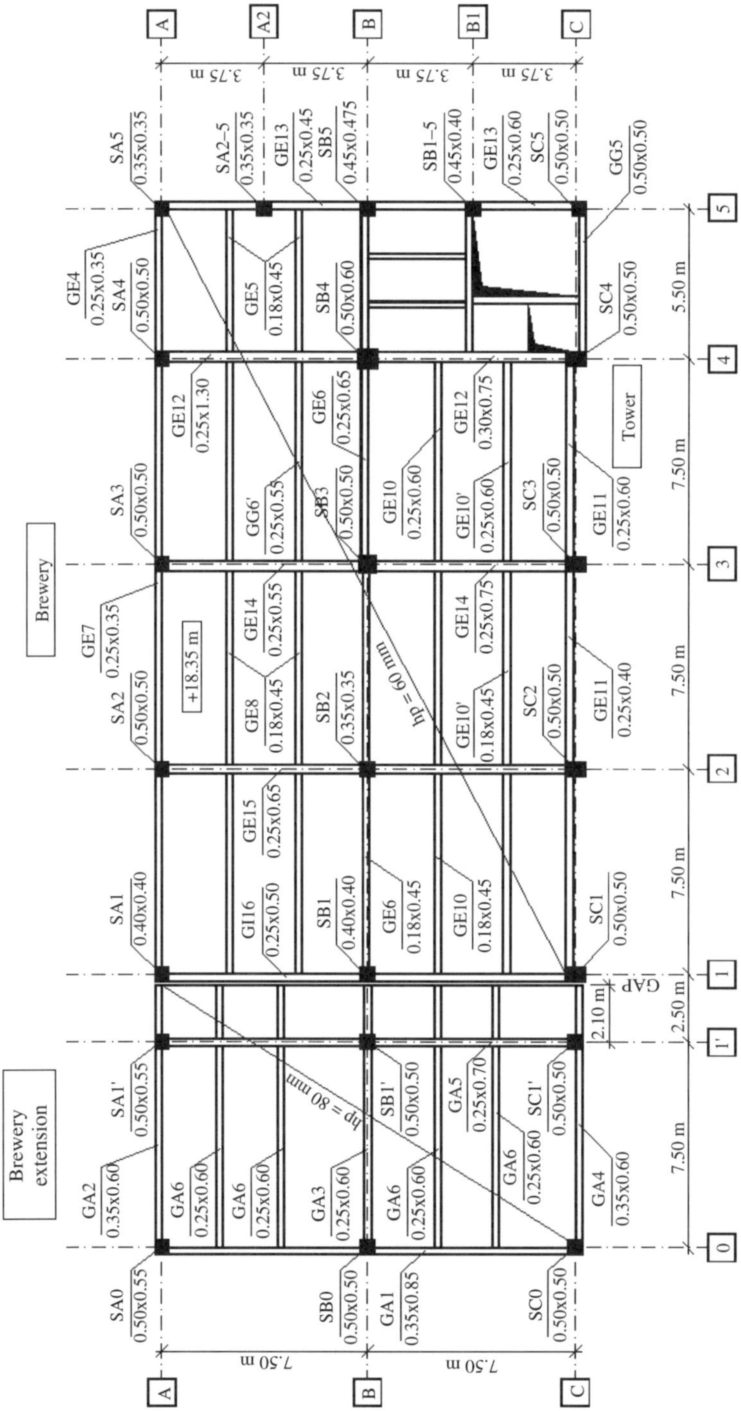

Fig. 5.4: The Timisoreana Brewery—framing plan, level +18.40 m

Fig. 5.5: Damage of secondary beams

Fig. 5.6: Damage of main beams

concrete cover spalled over a large surface; complete corrosion of many stirrups and deep corrosion of main reinforcement; some broken reinforcement. The damaged areas were located at the second floor (level +10.80 m), in the middle of the span for secondary beams (*Fig. 5.5*), on potential plastic hinge regions of the main girder (*Fig. 5.6*) and on columns (*Fig. 5.7*). Such damage was caused by the action of chloride ions (Cl^-) from salt solution, which was stored on the second floor, as well as by CO_2, RH $\approx$ 80% and temperatures over 40°C.

In some main beams, dangerous inclined cracks were also detected at the secondary beam to main beam connections (*Fig. 5.8*) due to inadequate transversal reinforcement at shear force.

Inadequate longitudinal reinforcement was deduced from the structural analysis. The initial analysis done in 1960 was performed according to Romanian codes [5] at low seismic design actions, owing to weakness in the structural system at present-day high seismic actions.

5.2.2 Assessment of *in situ* conditions

Non-destructive tests such as rebound test as well as pulse velocity measurements were performed on the main structural elements. The average values are presented in *Table 5.1*. The mean compressive strength (f_{cm}) of the investigated elements was obtained by using the combined method: pulse velocity (v)–rebound index (n). The concrete class given by the combined method was: C8/10–C16/20 at columns; C12/15–C25/30 at main girders; C8/10 at secondary beams.

The results of the non-destructive analysis emphasized some important conclusions: the concrete class of the columns is good for some of the elements; the concrete class in many beams and slabs is below the minimum necessary for reinforced concrete floors.

Fig. 5.7: Damage of columns

Fig. 5.8: Inclined cracks at main beams

The specific service conditions of the structural elements ($T = 40-60°C$, $RH = 70–80\%$, chloride ions present) during the 42 years lead to some significant damages.

Concrete carbonation and/or chloride ion penetration was checked by both procedures: theoretical analysis and experimental test. The theoretical values of the concrete carbonation/ion penetration were calculated according to Ref. [6] and are presented in *Table 5.2*. The experimental measurements were made by pH test and the results are, also, illustrated in *Table 5.2*.

The carbonation of the covering concrete created the conditions for the reinforcement corrosion: 30–60% of main reinforcement steel cross section was corroded at some elements like columns and beams (*Table 5.3*). Reinforcement steel characteristics of the existing structural members used in 1960 are Romanian ribbed bars PC52 ($f_{yk} = 350$ N/mm^2;

$f_{yd} = 300$ N/mm^2) for longitudinal reinforcement and Romanian plain bars OB37 ($f_{yk} = 245$ N/mm^2; $f_{yd} = 210$ N/mm^2) for stirrups [7].

Element	n	v (m/s)	f_{cm} (N/mm^2)	Concrete class
Column SC2	38.4	3644	20.60	C12/15
Column SB2	45.1	3717	28.20	C16/20
Column SC1	39.7	3829	26.30	C16/20
Column SA2	47.8	3077	17.70	C8/10
Column SB3	41.9	3774	27.60	C16/20
Main beam GE14	36.9	3194	19.25	C12/15
Main beam GE15	48.4	3940	36.10	C25/30
Secondary beam GE10	41.1	3280	18.40	C8/10
Secondary beam GE8	36.2	3256	13.60	C8/10
Slab	34.2	2373	6.20	<C4/5

Table 5.1: Non-destructive analysis of concrete class

Element	Theoretical	Experimental
Columns	33	5*–10*
Secondary beams	50	20*
Main beams	26	20–25

*Experimental measurements were influenced by the periodical sanitation (with mortars) of the elements; new mortar layers had higher pH.

Table 5.2: Carbonation depth (mm)

Storey	Element	Reinforcement characteristics					
		Initial		Measured		$\Delta\Phi = \frac{\Phi_0-\Phi_r}{\Phi_0}$	$\Delta A = \frac{A_0-A_r}{A_0}$
		Φ_0 (mm)	A_0(mm^2)	Φ_r (mm)	A_r (mm^2)	×100 (%)	×100 (%)
III	Column	25	491	17	227	32	54
	Column	25	491	20	314	20	36
V	Column	25	491	20.5	330	18	33
	Column	25	491	21	346	16	30
	Main girder	22	380	20	314	9	17
	Main girder	22	380	19	283	14	26

Table 5.3: Reinforcement corrosion of main bars in some structural elements

5.3 Different strategies considered for strengthening

5.3.1 General solutions for rehabilitation

Regarding the rehabilitation, the main solutions for the vertical irregularities consist of:

- strengthening of existing structural elements and/or the structural system by increasing the strength stiffness and ductility of the weak structural elements;
- stalling additional structural members.

For both solutions, it is necessary to avoid new stiffness discontinuities under lateral displacement. On the other hand, strengthening of vertical members at some levels may involve rehabilitation of the floors. In the case of horizontal structural irregularities, the aim of rehabilitation is to reduce the eccentricity between the centre of stiffness and the centre of mass: the result is decrease of torsion forces and displacements as well as an increase of the strength with respect to lateral actions. The common solution is to use new symmetrical walls. For irregularities of the geometric plan, the rehabilitation solution consists of the use of new walls and/or seismic joints. The rehabilitation solutions for general structural vulnerabilities are presented below.

In the case of indirect transfer of strong forces and horizontal members with large span/high loads, the classical solution is to use additional columns (vertical or inclined) for transferring the strong forces to the existing (or new) foundations. For weak columns (compared with the adjacent beams) column strengthening is necessary. The rehabilitation solutions adopted in the case of deterioration of building component parts depend on the structural material.

Owing to structural vulnerabilities and/or torsion effects, elements of the system may be subjected to different displacements and some damages may result. Special rehabilitation systems may be used such as adding abutments in directions of low stiffness and building of additional reinforced concrete (RC) walls.

5.3.2 Specific solutions for strengthening of reinforced concrete structures

Reinforced concrete structures are to be repaired and/or strengthened in cases when the general damage is limited, and demolished when the structural safety is greatly affected and the rehabilitation cost is very high [8–10].

Repairs are used for surface deterioration, cracks, damage resulting from casting defects, and reinforcement corrosion. The methods used for repairs are jacketing of damaged surfaces; infilling of cracks with usual mortar, epoxy resin or other polymers; and replacement or strengthening of damaged reinforcement.

Strengthening of reinforced concrete structures takes into account the increase of strength, stiffness, and ductility. In case of reinforced concrete framed structures, the increase in stiffness and ductility is to be achieved by jacketing of beams, columns, and joints. The jacketing is performed by reinforced concrete, steel profiles, carbon fibres, carbon fibre reinforced polymer (CFRP), etc. CFRP may be used for increasing ductility and slightly increasing the stiffness; see Ref. [11].

For reinforced concrete frame-wall structures the increase of bearing capacity is obtained by coating the core, the flange, and the coupling beam.

Sometimes it is necessary to transform the existing structure completely, especially for framed structures. In this case, special techniques such as steel bracing of reinforced concrete structures and infilling of frame openings with reinforced masonry or reinforced concrete are to be used.

5.3.3 Adopted solutions for strengthening at Timisoreana Brewery

The assessment performed in 1999 showed local damages at slabs, main girders, secondary beams, and columns. As previously presented, the damage consisted of concrete carbonation; concrete cover spalled over a large surface; complete corrosion of many stirrups and deep corrosion of main reinforcement; and some broken reinforcement. Also, inadequate longitudinal reinforcement was deduced from the structural analysis. The initial design, done in 1960, was performed according to Romanian codes in effect at that time with provisions of low seismic actions, owing to structural system weakness at present-day high seismic actions. The necessary rehabilitation of the reinforced concrete structure was adopted and performed for all types of damages. The main girders and secondary beams were strengthened by jacketing with reinforced concrete. The columns were strengthened for both local damage and inadequate reinforcement, by jacketing with reinforced concrete over two storeys. The existing foundation was jacketed over and around with reinforced concrete for secure fixing of new main reinforcement of the column.

In 2003, due to continuous operation and subsequent damage of the structure, a new assessment was required. It was found that some beams and one column were characterized by inadequate longitudinal reinforcement (in the column) and shear reinforcement as well as corrosion of many stirrups at beams. The necessary strengthening was performed at beams and column characterized by inadequate longitudinal reinforcement (the column) and shear reinforcement as well as corrosion of many stirrups (five beams). The strengthening solution adopted was based on CFRP composites.

5.4 Structural analysis before and after repair, design of sections, and codes

5.4.1 Structural analysis and response of structure to loads before and after repair

According to the Romanian codes of actions [12], the structural analysis was performed for the persistent and transient design situation and the accidental design situation by taking the seismic action into account at present-day magnitude [1]. The load characteristics are given in *Table 5.4*.

According to the Romanian Code for seismic design P100-92 [1] as well as other norms, the design of structures to resist earthquake is based on the following procedures and calculation methods.

- Common design procedures based on the following calculation methods: *linear static* with conventional forces distributed as inertia forces for linear static response; *linear dynamic* with accelerograms for modelling of seismic actions.
- Design procedure based on consideration of post-elastic deformation of structures with *non-linear static* analysis and conventional forces distributed as inertia forces for seismic response; *non-linear dynamic* method with accelerograms for modelling of seismic action.

Load case		**Dead load**	**Imposed load**	**Live load**	**Snow load**	**Wind load**	**Seismic load**
Characteristic load		5.0 kN/m^2	1.0–10.0 kN/m^2	2.0 kN/m^2	0.7 kN/m^2	0.7 kN/m^2	$a_s = 0.16\ g$ $\beta_{\max} = 2.5$ $q = 5.0$
Load factor	Persistent and transient design situation	1.2	1.2	1.3	0.7	1.2	0.0
	Accidental design situation	1.0	1.0	0.8	0.3	0.0	1.0

Table 5.4: Load cases and combinations

Assessment of the existing structures with respect to seismic action is estimated according to the Romanian code by calculus of the earthquake capacity ratio R:

$$R = \frac{S_{\text{cap}}}{S_{\text{nec}}} \tag{5.1}$$

where S_{cap}: seismic shear force capacity (seismic base shear force); S_{nec}: conventional seismic load (seismic base shear force) calculated according to the Romanian code P100-92 for seismic design action.

For the assessment of existing structures, the general Eq. (5.1) may be written for different sectional efforts and applied for individual structural members, as for instance:

$$R = \frac{M_{cap}}{M_{nec}} = \frac{M_{\text{Rd}}}{M_{\text{Ed}}} \tag{5.2}$$

where M_{cap} or M_{Rd}: resistance bending moment; M_{nec} or M_{Ed}: design bending moment calculated for the present-day level of actions. The equivalence between the Romanian earthquake capacity ratio and the more common safety approach according to EN 1990 [13] is presented in *Table 5.5*.

Building class of importance		**Earthquake capacity ratio R_{min}**	**Global safety coefficient C_0**	**Reliability index β**	**Failure probability P_{f}**
New buildings		1.00	2.250	4.75	10^{-6} to 10^{-7}
Existing buildings of class	I. Buildings of vital social importance	0.70	1.575	3.09	10^{-3}
	II. Very important buildings	0.60	1.350	2.00	2×10^{-2}
	III. Normal importance buildings	0.50	1.125	1.28	10^{-1}
	IV. Reduced importance buildings	0.50	1.125	1.28	10^{-1}

Values are given for normal distribution of actions and strengths and variation coefficient $C_{\text{v}}^{\text{r}} = C_{\text{v}}^{\text{a}} = 10\%$.

Table 5.5: Safety factors of new and existing buildings

The Timisoreana Brewery, an existing industrial building of normal importance (class III), has to satisfy the earthquake capacity ratio $R_{\text{min}} = 0.5$ corresponding to the failure probability $P_{\text{f}} = 10^{-1}$.

5.4.1.1 Advanced structural analysis

The authors used appropriate procedures based on consideration of post-elastic deformation with non-linear analysis for design. These procedures were used for analysis and redesign of existing structures in seismic regions [14]. For damage control of structural members at seismic design the authors proposed and used the stiffness modification procedure. The stiffness modification procedure [14] is based on the influence of stiffness degree calculated as function of material characteristics: elasticity modulus (E_s, E_c) and area (A_s, A_c) of reinforcing steel and concrete. For instance, according to the Romanian design code for reinforced concrete structures [7], at bending with/without axial force the stiffness is given by the formula:

$$K = \frac{E_s \times A_s \times \beta \times d^2}{1 \mp ([\xi - \overline{x}_s]/\overline{e}_0)} \tag{5.3}$$

where E_s: elasticity modulus of reinforcing steel; A_s: area of tension reinforcement; d: effective depth of reinforced concrete cross section; $\overline{e}_0 = e/d$: relative eccentricity of axial force N; $\overline{e}_0 = \infty$ for pure bending; $\overline{x}_s = x_s/d$ where x_s is the distance between reinforcement area A_s and centroid of the concrete cross section; $\beta = \zeta(1 - \xi)/\psi$ where $\xi = x/d$ and x is the depth of neutral axis; $\zeta = (d - x/2)/d = 1 - \xi/d = 1 - \xi/2$; ψ given in *Table 5.6*, see Ref. [7].

ν ratio between long term action and total action	**Reinforcement (%)**		
	0.2–0.5	**0.5–0.8**	**>0.8**
$\nu \leq 0.5$	0.8	0.9	1.0
$\nu < 0.5$	0.9	1.0	1.0

Table 5.6: Values of ψ [7]

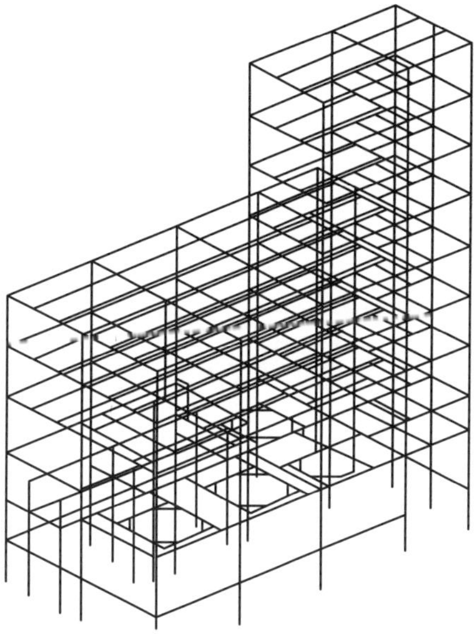

Fig. 5.9: Spatial structure analysed by FEM

Finite element method (FEM) analysis is used and it is possible to assign different values of stiffness K for each element. The procedural advantages arise from the opportunity to change the value of K at any time of RC structure utilization, for example, after serious degradation of one or several structural members.

5.4.1.2 Structural analysis carried out before repair

Inadequate longitudinal reinforcement was deduced from the structural analysis. The initial design done in 1960 was performed according to Romanian codes, under which the building's seismic design load had very low magnitude, owing to weakness in the structural system.

The actual structural analysis and assessment was performed at present-day level seismic action by FEM on the spatial structure (presented in *Fig. 5.9*).

In order to quantify the influence of structural damage in structural analysis, the stiffness modification procedure was used. Due to reinforcement corrosion of transversal beam GG11 and longitudinal beam GG8 (*Figs. 5.2 and 5.3*) reduced values of stiffness K, as given by Eq. (5.3), were taken into account for the structural analysis. The present structural analysis results quantified by the earthquake capacity ratio $R = M_{Rd}/M_{Ed}$ [Eq. (5.2)] for the damaged structure are presented in *Table 5.7.*

Element	**Transversal seismic action**				**Longitudinal seismic action**			
	N_{Ed} (kN)	M_{Ed} (kNm)	M_{Rd} (kNm)	$R = \frac{M_{Rd}}{M_{Ed}}$	N_{Ed} (kN)	M_{Ed} (kNm)	M_{Rd} (kNm)	$R = \frac{M_{Rd}}{M_{Ed}}$
SA2 ground storey	1696	145	285	1.96	1460	136	209	1.53
SA3 ground storey	1439	218	332	1.52	1400	202	216	1.07
SB3 ground storey	3907	1991	1748	0.88	3539	1918	1737	0.91
SB4 ground storey	4615	762	342	0.44	4643	524	342	0.65
SA2* storey I	1139	195	195	1.00	1179	174	159	0.91
SA3* storey I	1042	262	223	0.85	1332	209	160	0.76

*Resistance capacity was calculated with the diminished area of the main reinforcement.

Table 5.7: Analysis results for columns

From the structural analysis data presented it can be concluded:

- most of the actual values of earthquake capacity ratio $R < R_{min} = 0.50$;
- for column SB4 value of $R = 0.44$;
- low values of $R < 1.00$ were obtained for columns SB3, SA2, and SA3.

According to Romanian design codes for existing structures, when $R \leq R_{min} = 0.50$ for normal importance (class III) buildings, strengthening is necessary. A special analysis was performed on the floor beams GE14 and GE15 of the fourth storey (*Fig. 5.4*), where some inclined and dangerous cracks were present (*Fig. 5.8*), due to inadequate shear reinforcement (stirrups or/and inclined bars) near the force load (around the secondary beam) where the shear force has an important sensitive value.

Shear force resistance was calculated according to inclined cracks theory [7]. The fundamental condition for checking shear forces at the ultimate limit state is $V_{Rd}/V_{Ed} \geq 1$. From the data presented in *Table 5.8* it can be seen that all elements are vulnerable and a strengthening solution is necessary.

Storey	**Element**	**Design shear force V_{Ed} (kN)**		**Resistance shear force**	$\left(\frac{V_{Rd}}{V_{Ed}}\right)_{min}$
		Persistent and transient design situation	**Accidental design situation**	V_{Rd} **(kN)**	
III	Transversal main beam	270	313	206	0.66
	Longitudinal main beam	130	180	125	0.69
IV	Secondary beam	154	0	121	0.79
VI	Transversal main beam	10	138	93	0.67
	Transversal main beam	262	281	172	0.61

Table 5.8: Static and dynamic analysis results for beams at shear forces

5.4.1.3 Structural analysis carried out after repair

Several strengthening solutions were proposed and analysed. Structural redesign was performed by FEM on the spatial structure presented in *Fig. 5.9*, at present-day level actions. Initial rehabilitation of the reinforced concrete structure, performed in 1999, was adopted for both types of damages and consisted of reinforced concrete jacketing of beams, columns, and foundations. Due to the inadequate main reinforcement in columns SB3 and SB4 some strengthening solutions, by reinforced concrete jacketing, were studied with the results of structural analysis quantified by the earthquake capacity ratio $R = M_{Rd}/M_{Ed}$ [Eq. (5.2)] presented below (*Table 5.9).*

Element	**Transversal seismic action**			**Longitudinal seismic action**		
	M_{Ed} **(kNm)**	M_{Rd} **(kNm)**	$R = \frac{M_{Rd}}{M_{Ed}}$	M_{Ed} **(kNm)**	M_{Rd} **(kNm)**	$R = \frac{M_{Rd}}{M_{Ed}}$
Strengthening solution A: column SB3 from +4.40 to +10.80 m column SB4 from foundation to +10.80 m						
SB3 ground storey	376	356	0.95	335	278	0.83
SB4 ground storey	1585	1358	0.86	1193	1358	1.14
Strengthening solution B: column SB3 from foundation to +10.80 m						
SB3 ground storey	2113	2214	0.91	1855	1984	1.07
SB4 ground storey	594	326	0.55	456	321	0.70

Table 5.9: Redesign of strengthened structure: efficiency of different solutions

For both strengthening solutions $R > R_{min} = 0.50$ is necessary for existing buildings of class III (*Table 5.5*). Finally, due to economic reasons, the strengthening solution B, only for the column SB3, was chosen. The main girders and secondary beams were strengthened by coating with reinforced concrete. New longitudinal reinforcement bars and stirrups were located at the bottom of each beam in a new concrete layer of 150 mm depth. The column SB3 was strengthened for both local damage and inadequate reinforcement. The jacketing with reinforced concrete was used over two storeys and consists of 225 mm depth on all four sides. The existing foundation was jacketed over and around by 500 mm depth reinforced concrete for secure fixing of new main reinforcement of the column.

At the assessment performed in 2003, due to continuous operation and subsequent damage of the structure, it was found that some beams and one column were characterized by inadequate longitudinal reinforcement (in the column) and shear reinforcement as well as corrosion of many stirrups at beams. The strengthening solution adopted was based on CFRP. Structural analysis carried out after repair shown the same results as before repair since no cross section dimensions changes were performed.

5.4.2 Codes

The structural analysis was performed according to the Romanian codes of actions [12] by taking the seismic action into account at present-day magnitude [1]. The assessment of the existing structures with respect to the seismic action is done according to the Romanian code P100-92 [1] by calculus of the earthquake capacity ratio. The initial rehabilitation of the reinforced concrete existing structure was performed by jacketing with reinforced concrete. The analysis before and after strengthening was done according to the Romanian code for design and detailing of reinforced concrete structural members, STAS 10107/0 – 90 [7]. The final strengthening was based on CFRP composites. The design and detailing of strengthening solutions were done according to *fib* Bulletin 14 [15] and *fib* Bulletin 35 [16] for retrofitting of concrete structures by externally bonded CFRPs with emphasis on seismic applications. The cross section analysis after repair was done by using the SIKA Software with Sika CarboDur Composite Strengthening Systems to increase flexural, shear and confinement strength of reinforced concrete structures based on the *fib* Bulletin 14.

5.5 Detailing

The rehabilitation of the reinforced concrete structure adopted and performed in 1999 for both types of damages consisted of jacketing with reinforced concrete of deteriorated beams, one column and its foundation. The main girders and secondary beams were strengthened by coating with reinforced concrete. New 4ϕ 25 mm reinforcement bars for each secondary beam and 6ϕ 25 mm reinforcement bars for main girder were placed at 150 mm from the bottom side of the beams with new stirrups ϕ8/150 mm (*Figs. 5.10a, b and 5.11a*). One column

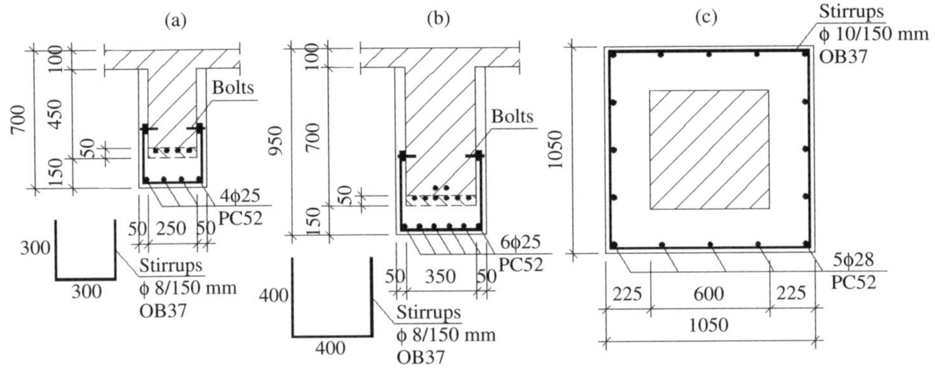

Fig. 5.10: RC jacketing solutions: (a) secondary beam; (b) main girder; (c) column (Units: mm)

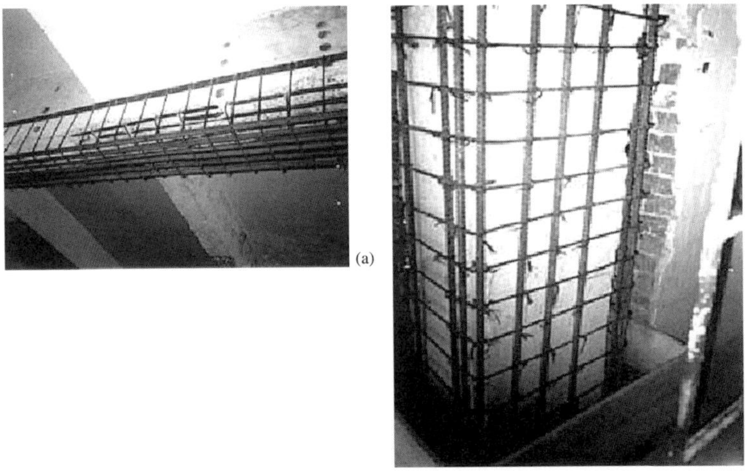

(a) (b)

Fig. 5.11: RC strengthening of: (a) main girder; (b) column

was strengthened for both local damage and inadequate existing reinforcement. The coating with reinforced concrete was used over two storeys and consists of 16ϕ28 mm longitudinal reinforcement bars, stirrups ϕ10/150 mm, and 225 mm concrete depth on all four sides (*Figs. 5.10c and 5.11b*). The existing foundation was jacketed over and around by 500 mm depth reinforced concrete for secure fixing of the column new main reinforcement. Reinforcement steel characteristics used for strengthening are Romanian ribbed bars PC52 ($f_{yk} = 350$ N/mm^2; $f_{yd} = 300$ N/mm^2) for longitudinal reinforcement and Romanian plain bars OB37 ($f_{yk} = 245$ N/mm^2; $f_{yd} = 210$ N/mm^2) for stirrups. New concrete was class C20/25.

The new longitudinal reinforcement bars from beams were anchored at the ends by welding on steel plates fixed in the nodes by steel collars around the end nodes of the existing concrete structure. The new stirrups from beams were welded on longitudinal continuous steel plates fixed in the web of existing beams by using mechanic bolts. All these detailing aspects will be further illustrated in *Section 5.6*, construction procedures.

The strengthening, performed in 2003, was used for some beams and one column characterized by inadequate flexural and shear reinforcement. The strengthening solution adopted was based on carbon fibre polymer composites (CFRP) as it is illustrated in *Figs. 5.12 and 5.13*. The column was strengthened at the ground storey by longitudinal Sika Carbodur S1012 strips on each side of 100 mm width and 1.2 mm thickness. The strips were placed in different position in the cross section to pass by the structure node. As shear strengthening, a single layer of Sika wrap HEX 230C closed jacket was used on 1.20 m height at the ends of the column. The sheets had 600 mm width and 0.12 mm thickness. The beams were strengthened at several stories by a longitudinal Sika Carbodur S1012 strip of 100 mm width and 1.2 mm thickness. The strips were placed at the bottom side of the cross section as necessary from design. As shear strengthening, a single layer of Sika wrap HEX 230C open jacket was used on 1.20 m height at the ends of the beams. The sheets had 600 mm width and 0.12 mm thickness. CFRP material characteristics used for strengthening are $E_f = 165$ kN/mm^2 and $\varepsilon_{fu} = 0.017$ for longitudinal strips; $E_f = 231$ kN/mm^2 and $\varepsilon_{fu} = 0.017$ for transversal wraps.

(a) (b)

Fig. 5.12: CFRP strengthening of: (a) column; (b) main girder (Units: mm)

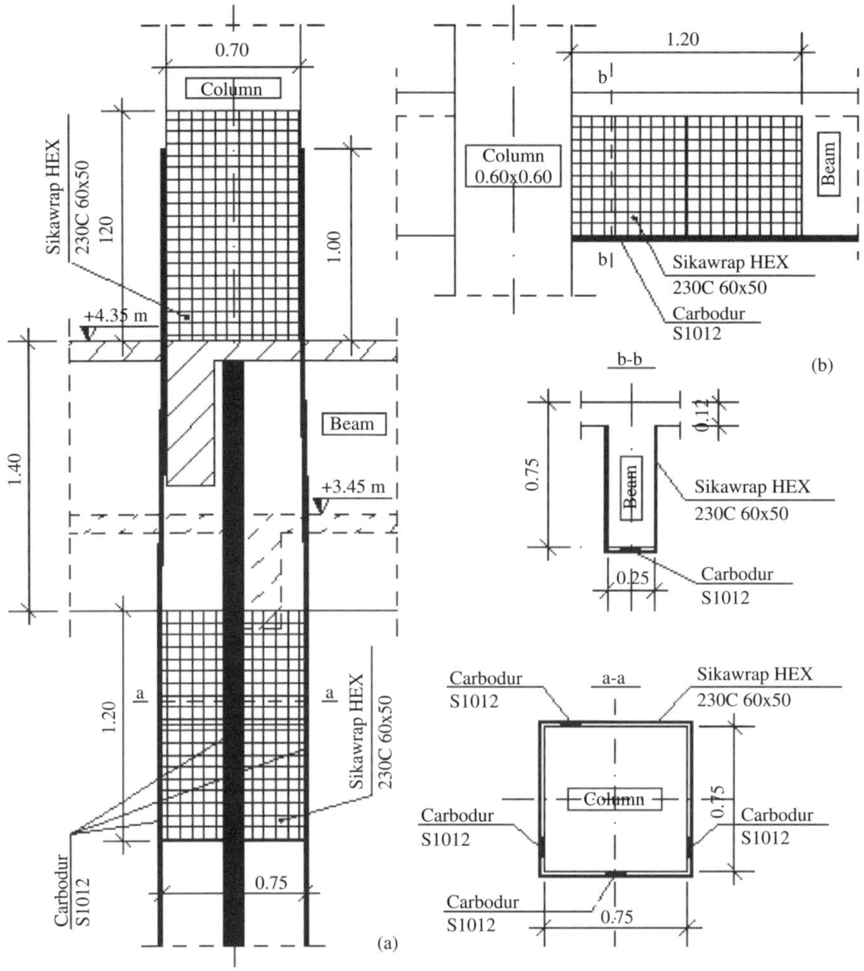

Fig. 5.13: CFRP strengthening details for: (a) column; (b) main girder (Units: m)

The bond of CFRP materials to the existing concrete layer was ensured by specific adhesives (Sikadur). The longitudinal CFRP strips for column strengthening were anchored in holes of 200 mm depth performed into existing reinforced concrete foundation. The longitudinal CFRP strips for beam strengthening started at the face of column-beam node as it were used as lower reinforcement in the beams span. Ordinary protection of CFRP strengthening materials was ensured by a cement mortar layer. All these detailing aspects will be further illustrated in *Section 5.6*, construction procedures.

5.6 Construction procedures

The first rehabilitation of the reinforced concrete structure was performed as follows. The beams were strengthened by coating with reinforced concrete at the bottom side for embedding the new longitudinal reinforcement bars and on the two lateral sides

Fig. 5.14: Anchorage detailing at beams end

Fig. 5.15: Reinforcement detailing at beams

Fig. 5.16: Reinforcement detailing at column–foundation joint

Fig. 5.17: Surface repair and preparation before CFRP application

for embedding the new stirrups. The column was strengthened by coating with reinforced concrete on all four sides. The construction steps for reinforced concrete jacketing were:

- Spalling of damaged concrete cover and mechanical cleaning (sand blasting) of existing concrete substrate.
- Fixing of steel collars around the beam-column nodes by mechanical bolts and the steel plates for welded anchorages of longitudinal new reinforcement bars from beams (*Fig. 5.14*).
- Placing of longitudinal new reinforcement for beams (*Fig. 5.15*) and for columns with the secure fixing into foundation (*Fig. 5.16*).
- Placing of the new transversal stirrups for beams and columns. The stirrups from beams were welded on longitudinal continuous steel plates fixed in the in the web of existing beams by using mechanic bolts.
- Manufacturing and placing of timber framework and shoring.
- Casting of concrete.

The second strengthening was performed as follows. The column was strengthened with longitudinal CFRP strips on all four sides. The strips were placed in a different position in the cross section to pass by the structure node. As shear strengthening, a single layer of CFRP closed jacket sheet was used at the ends of the column. The beams were strengthened by a longitudinal CFRP strip placed at the bottom side of the cross section. As shear strengthening, a single layer of CFRP open jacket sheet was used at the ends of the beams. The construction steps for CFRP strengthening were:

- Mechanical cleaning (sand blasting) of existing concrete substrate and dust removal.
- Sealing of existing cracks and surface repair with Sika epoxy-based mortars for obtaining a smooth plane application surface (*Fig. 5.17*).
- Mechanical rounding of concrete cross section corners at 20 mm radius for CFRP sheet application.

Fig. 5.18: End anchorage of longitudinal CFRP strips for column

- Application of specific two components mixed adhesives on the concrete substrate. The used adhesives were Sikadur-30 for CFRP longitudinal strips and Sikadur-330 for CFRP sheets.
- Placing CFRP strips (*Figs. 5.19 and 5.20*) bonded by previously presented adhesives. The end anchorage of longitudinal CFRP strips for columns was ensured into holes of 200 mm depth performed into existing reinforced concrete foundation (*Fig. 5.18*). The longitudinal CFRP strips for beams started at the face of column-beam node as it was used as lower reinforcement in the beams span.
- Placing of CFRP sheets (*Figs. 5.19 and 5.20*) bonded by previously presented adhesives. The CFRP wrapping was applied on four sides for column as closed jacket with a horizontal overlapping of 100 mm and on three sides for beams as open jacket.
- Protection of CFRP strengthening materials by a cement mortar layer.

(a) (b)

Fig. 5.19: CFRP application for column: (a) longitudinal strips; (b) transversal sheets

(a) (b)

Fig. 5.20: CFRP application for beams: (a) longitudinal strips; (b) transversal sheets

5.7 Summary

Many structures built in Romania before 1970 were designed for gravity loads with inadequate lateral load resistance because earlier codes specified lower levels of seismic loads. Some of these structures are still in service beyond their design life. Moreover, some deterioration of component parts of buildings is encountered in existing structures due to the actions of different hazard factors.

The paper presents a case study of a brewery reinforced concrete framed structure of five storeys and a tower of nine storeys, which has been assessed and strengthened. The brewery and the tower were built in 1961 and an extension in 1971. The main problems comprised local damage of some structural elements caused by the action of carbon dioxide and/or chloride ions and inadequate reinforcement of columns and beams at seismic actions.

The assessment performed in 1999 showed up local damages at slabs, main girders, secondary beams, and columns. The damage consisted of concrete carbonation, concrete cover spalled over a large surface, complete corrosion of many stirrups and deep corrosion of main reinforcement, and some broken reinforcement. Such damage was caused by the action of chloride ions (Cl^-) from salt solution, which was stored on the second floor as well as of CO_2, RH $\approx$ 80%, and temperatures over 40°C. Moreover, inadequate longitudinal reinforcement was deduced from the structural analysis. The initial design, done in 1960, was according to Romanian codes of that time with provisions at low seismic actions, owing to structural system weakness at present-day high seismic actions. The necessary rehabilitation of the reinforced concrete structure was adopted and performed as jacketing with reinforced concrete for main girders, secondary beams, and columns. The existing foundation was jacketed over and around by 500 mm depth reinforced concrete for secure fixing of the column new main reinforcement. In 2003, due to continuous operation and subsequent damage of the structure, a new assessment was required. It was found that some beams and one column were characterized by inadequate longitudinal reinforcement and shear reinforcement as well as corrosion of many stirrups at beams. The strengthening solution adopted was based on CFRP composites for beams and column.

The authors used for design the proper procedures based on consideration of post-elastic deformation with non-linear analysis. These procedures were used for analysis and redesign of existing structures in seismic regions. For the damage control of structural members at seismic design, the authors proposed and used the stiffness modification procedure.

The strengthening solutions proposed and analysed for the existing industrial building of normal importance class, Timisoreana Brewery, were adopted and erected to satisfy the structural behaviour demands at present-day level of actions, according to the Romanian design codes.

5.8 Acknowledgements

The current paper is the result of a five year period of assessment, design, and construction which took place at Timisoreana Brewery, Timisoara, Romania. The authors would like to express their gratitude to Civ. Eng. G. Rogobete from the brewery staff for his continuous help and technical support during the entire duration of the work on site. The authors are also grateful to Dr Civ. Eng. C. Suma for his important contribution in the specific strengthening project.

References

[1] Romanian Ministry of Public Works and Territory Planning. *Code for Aseismic Design of Buildings, Agrozootechnical and Industrial Structures, P100-92, English Version.* Romanian Ministry of Public Works and Territory Planning, Bucharest, 1992.

[2] Technical University of Civil Engineering Bucharest. *Code for Assessment and Strengthening Design of Existing Buildings to Seismic Action, P100-3 Project (in Romanian).* Technical University of Civil Engineering Bucharest, Bucharest, 2003.

[3] Ghobarah A. Seismic assessment of existing RC structures. *Prog. Struct. Eng. Mater.* 2000; **2**(1), pp. 60–71.

[4] Kölz E, Bürge M. Priorities in earthquake upgrading of existing structures. *Struct. Eng. Int.* 2001; **11**(3), pp. 202–206.

[5] Romanian Ministry of Public Works and Territory Planning. *Code for Design of Buildings and Industrial Structures in Seismic Regions, P.13-63 (in Romanian).* Romanian Ministry of Public Works and Territory Planning, Bucharest, 1963.

[6] Bob C. Some aspects concerning corrosion of reinforcement. *Proceedings of the International Conference "The Protection of Concrete"*, Dundee, E & FN SPON, London, September 1990.

[7] Romanian Ministry of Public Works and Territory Planning. *Design and Detailing of Concrete, Reinforced Concrete and Prestressed Concrete Structural Members, STAS 10107/0-90 (in Romanian).* Romanian Ministry of Public Works and Territory Planning, Bucharest, 1990.

[8] Packer JA. Tubular construction. *Prog. Struct. Eng. Mater.* 2000; **2**(1), pp. 41–49.

[9] Paulay T. A re-definition of the stiffness of reinforcement concrete elements and its implications in seismic design. *Struct. Eng. Int.* 2001; **11**(1), pp. 53–60.

[10] Park R, Paulay T. *Reinforced Concrete Structures*. Wiley, New York, 1975.

[11] Dan S, Bob C, Gruin A, Badea C, Iures L. Strengthening of reinforced concrete framed structures in seismic zones by using CFRP. *WSEAS Conference "Engineering Mechanics, Structures, Engineering Geology (EMESEG '08)"*, Heraklion, Crete, July 2008.

[12] Romanian Ministry of Public Works and Territory Planning. *Actions in Civil Engineering. Load Cases and Combinations for Buildings and Industrial Structures, STAS 10101/0A-77 (in Romanian)*. Romanian Ministry of Public Works and Territory Planning, Bucharest, 1977.

[13] European Committee for Standardization. *Eurocode 0: Basis of Structural Design, EN1990*. European Committee for Standardization, Brussels.

[14] Dan S, Bob C, Iures L. Assessment and redesign of RC existing structures in seismic regions. *Second International fib Congress*, Naples, June 2006.

[15] International Federation for Structural Concrete. *fib Technical Report: Externally Bonded FRP Reinforcement for RC Structures—Bulletin 14*. International Federation for Structural Concrete, Lausanne, October 2001.

[16] International Federation for Structural Concrete. *fib Technical Report: Retrofitting of Concrete Structures by Externally Bonded FRPs with Emphasis on Seismic Applications—Bulletin 35*. International Federation for Structural Concrete, Lausanne, April 2006.

[17] Bob C. Rehabilitation of existing structures in seismic zones. *Prog. Struct. Eng. Mater.* 2001; **3**(3), pp. 353–359.

[18] Bob C, Dan S. Repair and renewal of existing structures in seismic zones. *IABSE Symposium "Metropolitan Habitats and Infrastructure"*, Shanghai, September 2004.

[19] Bob C, Dan S, Badea C, Iures L. Classic and modern rehabilitation techniques in seismic zones. *IABSE Symposium "Structures and Extreme Events"*, Lisbon, September 2005.

Chapter

6

Strengthening and Rehabilitation of a Heating Plant Chimney, in Poland

Andrzej B. Ajdukiewicz, Prof., Dr; **Jacek S. Hulimka**, Assist. Prof., Dr
Silesian University of Technology, Gliwice, Poland

Abstract: A case study of reinforced concrete chimney repair, strengthening, and finally general modernization is described. The specific local conditions and the changeable decisions of the user caused the application of three different approaches to the reconstruction works.

Keywords: concrete destruction; construction faults; repair methods; reinforced concrete chimney; modernization processes; advanced strengthening methods.

6.1 Introduction

This paper describes a case study of a reinforced concrete (RC) chimney structure, over 80 m high, which has been repaired and reconstructed in a series of actions, particularly in three distinct stages.

The history of the RC chimney in the town heating plant goes back to 1976 when it was erected. From the beginning it had several defects. The chimney is situated in a seaside resort and serves as the only heating plant in the vicinity. The heating plant is responsible for heat and hot water supplies for the town of 45 000 people. The town includes a big spa district.

The chimney was designed and erected as a cylindrical shaped, RC structure of 80.5 m height. The outer diameter was 4.16 m and the inner diameter was 3.20 m. The total thickness of the wall consisted of a RC wall of the chimney carrying shaft 0.22 m thick, a heat insulation layer 0.12 m made of granulated slag, and a constructional reinforced wall of an exhaust gas conduit made of refractory concrete 0.14 m in thickness (*Fig. 6.1*). Both walls were erected simultaneously in double slip-shuttering. This detail influenced the quality of construction. The time of concrete setting in both shells was of particular importance. The contractor of the structure adjusted the speed of the boarding slide to the setting conditions of the refractory concrete in the inner shell. As a result, the concrete in the outer carrying shell showed numerous defects.

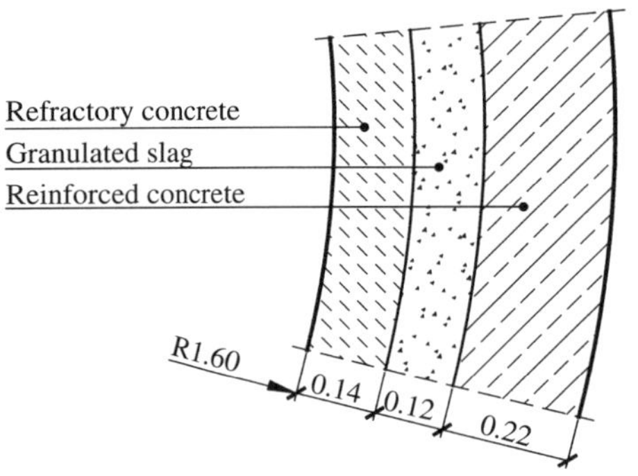

Fig. 6.1: Original cross section of the wall

The chimney was founded on a circular plate 13.0 m in diameter and of variable thickness from 1.6 m in the middle to 0.8 m on the edge. Taking into consideration the ground conditions (very moist layer of warp), the plate was placed on 39 cast *in situ* piles, each 12.0 m long.

The load-carrying shell was designed from concrete grade 200, which was common at that time (approximately C16/20, according to Eurocodes). It was to be reinforced with horizontal and vertical bars made of ordinary plain steel ($f_y \cong 250$ MPa). Designed circumferential reinforcement was of ∅10 mm bars in equal spacing of 0.17 m, whereas vertical reinforcement was designed in equal spacing of 0.15 m. The vertical reinforcement had changeable diameters, from ∅25 mm at the bottom, through ∅20 mm, and ∅16 to ∅12 mm over the level of +30 m.

The exhaust gas was led by means of a single steel smoke conduit. According to the functional assumptions, the chimney was to serve (simultaneously) five boilers of power rating 10 MW each and four boilers of power rating 5 MW each.

6.2 Early faults and series of emergency repairs

During the first years of utilization, the chimney was repaired and strengthened several times. As early as the first year of exploitation, it showed several significant defects. Numerous concrete defects seen on the outer surface of the load-carrying shell were the most vital. Defect concentration in the lower part of the chimney shaft was so severe that this part of the chimney was strengthened with steel bands (*Fig. 6.2a*). At the same time, the concept of shortening the chimney by 10 m was taken into consideration but due to ecological reasons, it was left as designed and erected.

After about ten years, due to strong corrosion of wrongly placed and compacted concrete, intensified by chloride attack from the sea winds and appearance of thermal cracks (up to 20 mm locally), the steel bands were taken off. At the same time, a strong outer jacket of RC, 0.12 m thick, was constructed to the height of about 40 m (*Fig. 6.2b*). Simultaneously, three massive steel bands were installed around the cleanout hole and the smoke conduit hole. The upper part of the chimney that was severely cracked as a result of temperature changes was strengthened with steel bands once more. The cracks were filled with bituminous substance for sealing.

During the next seven years, the structure was inspected occasionally; however, its user did not renovate the chimney as was suggested. The survey that was carried out at that time showed a local destruction at the level of +40 m. It was approximately the place where the additional RC jacket ended.

In such a state, the chimney survived to the year 1994 when the authors of this paper were commissioned to give an intricate expert opinion.

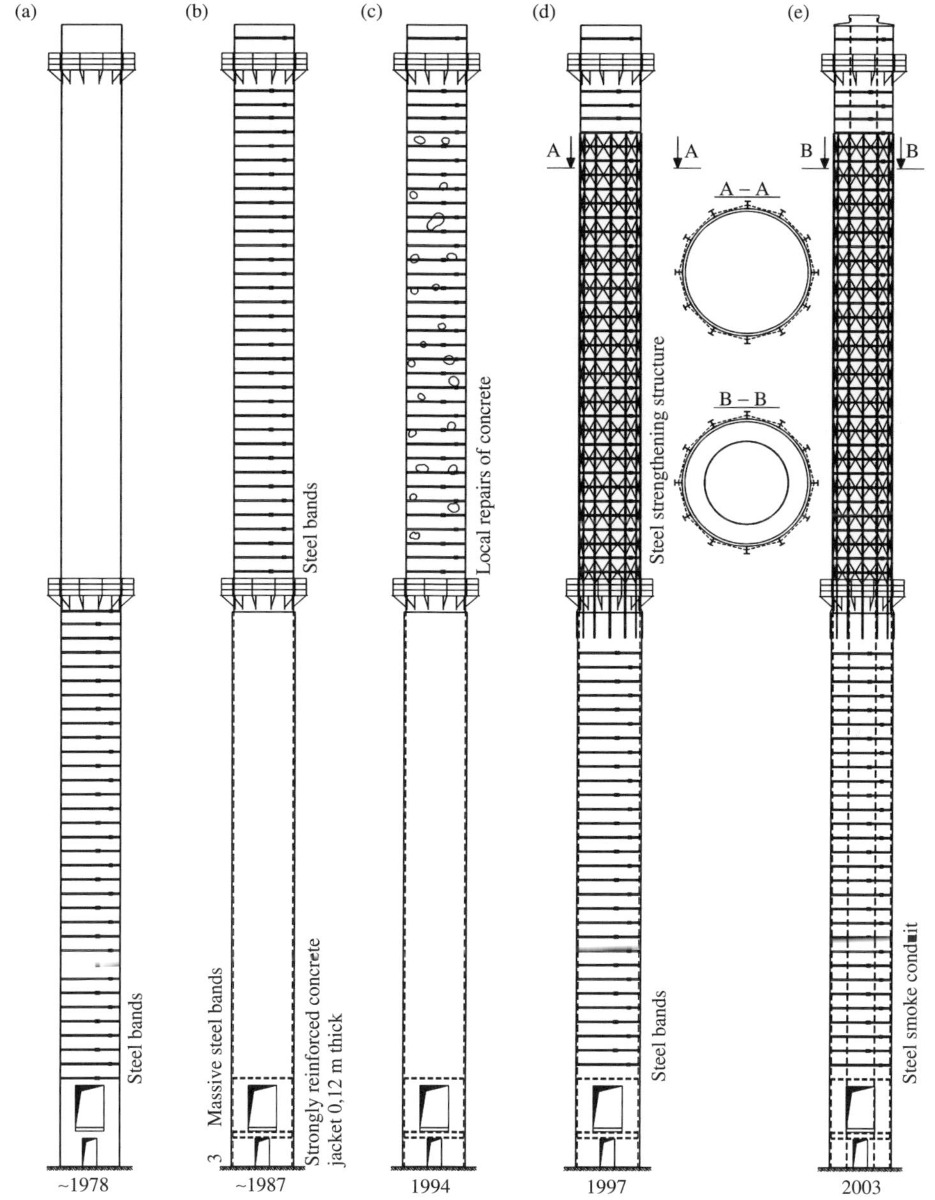

Fig. 6.2: Rehabilitation phases

The results of preliminary examination performed by the authors in 1994 showed such wide range of construction defects of the upper part of the shaft that the structure's condition was recognized as a pre-emergency one. At the time the following was recorded:

- numerous, considerably wide cracks up to 10 mm; in the past some cracks were filled with unknown bituminous substance which flowed out in a considerable amount under the influence of temperature;

Fig. 6.3: An example of outer jacket damage

- several concrete defects of various intensity; locally, they appeared across the whole carrying jacket (*Fig. 6.3*);
- local lack of concrete on the surface of over 0.5 m^2, filled with a wall of ceramic brick and masked with mortar;
- considerable reduction of reinforcement cross section in relation to design specifications; setting and binding defects of some reinforcing bars;
- penetration of combustion gases outside the shaft, which indicated dangerous discontinuity of the inner smoke conduit that was made of refractory concrete.

Material evaluation that was carried out during the first examination allowed to qualify some parts of concrete in the upper (accessible) part of the original shaft, which had maximum compressive strength of 12 MPa (C8/10 class) with pH factor not exceeding the value of 5.5.

In 1995 the chimney user lengthened the period of its exploitation to one additional year. And then, in 1996, it was decided that it was necessary to use the chimney for the next five years. During the described period, some additional examination of the inner part of the chimney was performed. It showed several serious defects of the inner jacket (*Fig. 6.4*) and complete lack of slag insulation that was specified in the original design.

6.3 Different strategies considered for repair

The damage state of the structure observed in 1994 was the result of various simultaneous factors, and in particular:

- careless original erection of the carrying shaft construction that was described above;

Fig. 6.4: An example of inner jacket damage

- lack of proper thermal insulation which resulted in water vapour outdropping with aggressive acid solutions out of exhaust gas;
- direct exhaust gas penetrating into the concrete through numerous leaks and structural defects, as well as through thermal cracks;
- exposure of the concrete to infiltration of water from the atmosphere and to chloride attack from the sea breeze;
- periodic exhaust gas influence on the surface of the shaft; the exhaust gas came out of the neighbouring chimney 40 m high.

In 1994, the structure's owner planned to use the chimney for no more than one or two heating seasons. Because of that, no attempts of comprehensive renovation were made. Recommendations were limited to necessary local repairs of the most important defects (*Fig. 6.2c*) and suggested regular monitoring of the structure's deflection. At the same time, a preliminary design of the chimney demolition was created. The chimney was to be replaced with a steel structure situated on the existing pile foundation. The necessity of chimney replacement in the time frame of about six months was an important design constraint. It was necessary because the town had to be supplied with heat.

The minimum five-year period of further exploitation that was imposed by the user in 1996 required strengthening in the upper part of the shaft. Such a need arose because of both observed concrete and reinforcement defects and the more distinct tilt of the upper part of the shaft. After thorough examination of possible strengthening methods, it was finally decided to design and build a spatial steel structure resting on top of the additional outer

RC jacket (*Fig. 6.2d*). The structure was to be anchored in the upper part of the jacket. The design assumed the transmission of all forces resulting from winds by a new structure to the lower part of the shaft strengthened with the additional jacket. The capacity of the pile foundation was one of the most important factors forcing the application of relatively light steel strengthening. Calculations showed the lack of capacity reserves preventing the building of additional load-carrying concrete jacket. The fact, in the past, that an additional concrete jacket was constructed was also taken into account in the calculations.

In 2001, after four years of failure-free utilization of the strengthened chimney, the owner of the structure decided to modernize the whole heating plant and to prolong its operation at least until 2015. Thorough modernization of all boilers was one of the stages. It was accompanied by installing dust collecting devices and additional heat exchangers. It resulted in considerable emission and exhaust gas decrease. Simultaneously, the temperature of exhaust gas at the chimney intake was lowered considerably.

It must be mentioned here that too large a diameter of the smoke conduit was one of the basic original defects of chimney construction. Such a large diameter of the smoke conduit was to match possible maximum exhaust gas expenditure during full load work of all boilers. The above-mentioned defect was typical for most chimneys serving several boilers built in 1970s and 1980s. Because of that and the total lack of thermal insulation that was in the original design, slow flowing exhaust gas was cooled considerably, which resulted in intensive outdropping of steam with sulphur and nitrogen compounds that are harmful for concrete and steel. The remains of such outdropping were visible on the outer surface of the smoke conduit. The location of the most intensive outdropping was in the middle of the chimney height corresponding to the place of the most intensive corrosive damages.

The anticipated exhaust gas cooling at the chimney intake would increase the steam outdropping and would lead to quick destruction of considerably damaged inner jacket made of heat-resistant concrete. The situation was serious because of numerous leaks in the inner smoke conduit. The leaks would enable the steam with acid solutions to migrate to the empty space between the jackets and it would lead to the destruction of the inner surface of the carrying jacket, which had been inaccessible so far. The authors had access to the measurements that were taken at the exhaust gas outlet of the first modernized boilers and the parameters showed that the chimney would loose the ability of proper exhaust gas transportation after the modernization of all boilers.

Under such circumstances, two basic possibilities were being considered:

- building several exhaust gas conduits into the inner part of the chimney, each one serving one boiler; from the point of view of the exhaust gas flow conditions such a solution would be the best, but after recalculating the necessary diameters of the conduits, it turned out that there was not enough room for them inside the RC construction;
- building a single smoke conduit inside the chimney; its parameters should be adapted to the possible expenditure and to the parameters of exhaust gas.

Because of the above-mentioned reasons, the single conduit variant was chosen. At the same time, it was decided to demolish the existing exhaust gas conduit made of refractory concrete. The decision was due to the poor condition of the inner surface of the carrying shaft, which had been inaccessible so far and had been exposed to intensive corrosive factors over many years.

6.4 Detailing and construction procedures at repairs

The repair works undertaken in 1994–1995 were provided according to the assumption of relatively short, not more than two-year use of the chimney. Thus, the range of repairs was limited to the main faults of the concrete surface of the load-carrying shell. These tentative repairs were performed with epoxy grouts or commercial cement–polymer mortars (PCC).

Fig. 6.5: The construction of steel strengthening of the upper part of the shaft

Following the decision of prolonged use of the chimney, the strengthening steel structure was designed and constructed in 1997. This strengthening spatial system covered all the upper part of the chimney, between the levels +40 m and +80 m, approximately.

The steel construction was designed as a system of 12 columns that were placed along the upper part of the shaft, over the additional RC jacket. The columns were strengthened with several intersecting tension members (*Fig. 6.5*) and were fixed to a steel ring (*Fig. 6.6*). The ring rested on the top of the additional outer RC jacket. The ring anchorage in a new jacket was obtained by leading over its surface short posts that were stabilized by numerous inserted anchors (*Fig. 6.7*).

Because of several cracks in RC outer jacket, steel bands on the lower part of the chimney were designed and installed.

The above-mentioned steel ring with anchorage elements was the most important of all structural elements because it was responsible for transmission of the reactions from the strengthening columns to the outer additional RC jacket. A typical ring element is schematically shown in *Fig. 6.8*.

All the elements of the strengthening were constructed in the summer of 1997. Because of the problems related to placing a suitably high crane at the building site, all the work was performed by mountain-climbing methods (*Fig. 6.9*). Such a system forced the strengthening structure to have been divided into elements of mass not more than 200 kg. Individual fitting-up elements were welded, whereas all fitting-up joints were bolted.

The visible damage of the outer jacket was repaired simultaneously with the fitting-up work. System mortars of PCC type were used for the repair work. Similarly, all important damages of the inner jacket that was accessible from the inner part of the exhaust gas conduit were repaired with the use of appropriately modified mortars.

The repaired and strengthened structure was admitted for use for about five years under the condition that it would be serviced every year and a survey would precede the servicing. Limit on the exploitation period was mainly due to the lack of access to the space between

Fig. 6.6: The detail of resting the steel members on the additional RC jacket

the jackets. This was the place where substantial corrosion of concrete and reinforcing steel was expected.

The decision for general modernization of the heating plant was the reason of the reconstruction process of the chimney to allow its usage up to 2015 at least.

Between 2002 and 2003, a design of smoke conduit demolition, a variant design of shaft's inner surface repair (depending on present technical condition), and a design of a new steel exhaust gas conduit were made (*Fig. 6.2e*). The design work was performed in 2003, between heating periods. The design assumptions included building an exhaust gas conduit as a self-supporting internal steel structure that was founded on the existing foundation plate. The structure was to be stabilized by slidable bearings to the shaft's inner surface and was to be built 0.5 m higher than the shaft top. The conduit diameter and its insulation (mineral wool and reinforced metal foil pinned onto the steel shell) were calculated according to the measurement results of exhaust gas parameters taken from the modernized boilers. Average heat need determined by the user's experience was taken into account. The whole structure was welded; the joint between the penultimate and ultimate segments was the exception to the rule because of the steel grade change; therefore, a collar bolted joint was made. While designing the smoke conduit, a 25 year exploitation period was assumed. The basic elements of the conduit construction are shown in *Figs. 6.10, 6.11, and 6.12*. The upper segment of the conduit was made of acid-resisting steel and all the other segments were made of stainless steel with surface protection paint.

Fig. 6.7: The detail of anchoring the steel members to the outer RC jacket

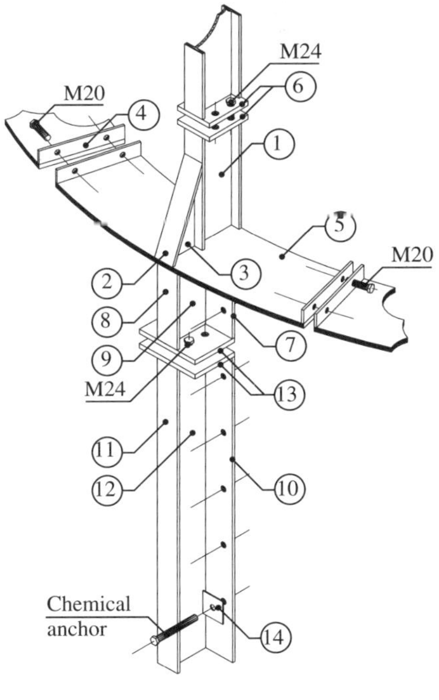

1	I 200/500
2	▭ 12/90x387
3	▭ 16/110x337
4	▭ 10/60x350
5	▭ 20/419x1243
6	▭ 20/160x200
7	▭ 12/280x355
8	▭ 12/100x355
9	▭ 10/167x355
10	▭ 12/280/1355
11	▭ 12/100x1355
12	▭ 10/176x1355
13	▭ 20/200x280
14	▭ 5/100x100

Fig. 6.8: Support element detail on the outer RC jacket

In order to stabilize the smoke conduit, two steel platforms were constructed (level +35.128 and +75.827 m). The platforms were anchored to the RC shaft. Four guide rolls with spring clamp were installed onto each platform. Steel guides were welded in suitable fragments of the conduit. They enabled free thermal deformation of the shaft. They were attached through the rollers (*Fig. 6.13*).

The smoke conduit demolition was executed carefully by cutting with diamond saws. The work was performed from a specially designed movable platform hanging from steel beams. The beams were stabilized on the shaft top.

Fig. 6.9: Assembly of the elements of the shaft steel strengthening

Fig. 6.10: Lower element of the smoke conduit during placing the thermal insulation

Complex repair of the inner surface of the carrying jacket, which was uncovered after the demolition, was one of the most important elements of reconstruction. Having finished the demolition, technical conditions of the inner surface of the load-carrying shaft were examined thoroughly. The examination showed several serious damages of concrete and reinforcement.

Fig. 6.11: Smoke conduit typical element installation

The concrete surface was eroded deeply, with numerous original defects such as improper concrete structure (*Fig. 6.14*) and the cover that was broken by slip-shuttering. In the places where the cover was insufficient, there was high corrosion of uncovered reinforcement bars (*Fig. 6.15*). In extreme cases, total destruction of individual bars was observed. The examination also confirmed considerable concrete decrements that were filled with a wall made of ceramic bricks (*Fig. 6.16*).

Taking into account the number and the character of observed damages, it was decided to sand-blast the whole shaft inner surface and, after completion or exchange of damaged reinforcing bars, to reconstruct the cover by shotcrete technology.

A ladder with a protection rail and lighting was designed and made inside the shaft.

Fig. 6.12: Shaft top element installation

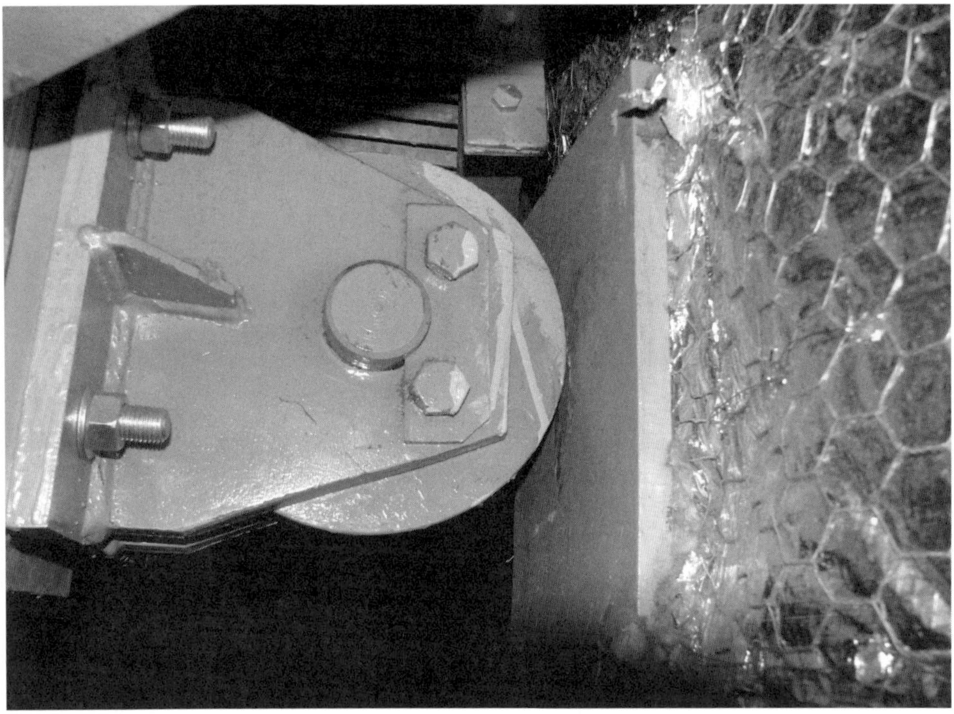

Fig. 6.13: A roller stabilizing the smoke conduit in the level of intermediate platform

Fig. 6.14: An example of concrete surface damages on the inner surface of the carrying shaft

Fig. 6.15: Shaft's inner surface; an example of uncovered and corroded reinforcement

Directly after the decision of the chimney complex modernization in the year 2001, the heating plant owner ordered to clean the whole shaft outer surface and steel strengthening and then to do small local repairs and to add paint protective layers onto the concrete and steel surfaces.

6.5 Current investigation results

In the years following the repair, a thorough examination of the whole RC and steel structure was made. This was made together with the examination of the inner surface of the new smoke conduit. The technical condition of all the elements was determined as good. The user's measurements of exhaust gas flow parameters confirmed that the assumptions, calculations, and construction solutions were correct. Yearly measurements of the shaft show the movement stabilization of its tilt. They can be compared with the ones taken directly after the construction of steel strengthening was made.

In the above situation, the chimney was allowed for operation for the period of forthcoming ten years. A condition was also specified that measurements and control would be performed periodically.

6.6 Summary

The presented case of a long-term repair, reconstruction, and modernization process of the RC chimney in the heating plant was forced by local conditions that made it impossible to

Fig. 6.16: Shaft's inner surface faults; visible void fillings with ceramic brick (outside masked with cement plaster)

demolish and reconstruct the object. As there was no decision concerning the period of the object exploitation, three fundamental stages of work had to be determined. First, they were emergency repairs on the assumption that the exploitation period would last no longer than two years. Second, the strengthening of the load-carrying structure was performed; assuming that the exploitation would last for the next five years. Lastly, assuming that it would be necessary to use the object for over ten years, a complex reconstruction with adaptation to important parameter changes of the exhaust gas source was performed.

Despite the necessity of adaptation to the user's various decisions and despite limiting the construction period for a maximum of six months between heating seasons, the object that was finally obtained fulfilled all the exploitation requirements and it should work without breaks in the forecasted period of about ten years.

The final rehabilitation works were assessed at about 25% of the replacement, and the period of these works was significantly shorter than that for construction of a new chimney.

Chapter

7

Rehabilitation of the Kumho Group Seoul Headquarters, Korea

Geonho Hong, Professor, Department of Architectural Engineering, Hoseo University, Asan, Chungcheongnam-do, Korea; **Youngsoo Chung**, Professor, Department of Civil Engineering, Chung-Ang University, Seoul; Anseong, Korea and **Hyekyo Chung**, CEO, DnK Construction Inc., Seoul, Korea

Abstract: This paper is a case study of an office building rehabilitation in Seoul, Korea. The partly built building, originally designed as a general office building, contained 20 stories above and seven below ground. After the first floor slab was constructed, construction was stopped because of financial difficulties of the previous owner. The new owner revised the architectural plan, design, and height of the building with 29 stories above and eight below ground. Because of the long-term stop of the construction and change of the architectural design, large-scale repair and rehabilitation work was carried out in 2006.

Keywords: rehabilitation; repair; office building; case study; demolition; extension.

7.1 Introduction

The owner of this building is a fully accredited and prosperous company involved in several business fields in Korea and worldwide. The company needed a new building for its second headquarters for their expanded personnel and department. After considering several buildings and locations, the company purchased this building in Chungro-Gu, Seoul.

The building, originally designed as a general office building, contained 20 stories above and seven below ground. After the basements and the first floor slab were constructed, construction was stopped in 1993 because of the financial difficulties (*Fig. 7.1*).

After purchasing the building, the owner revised the architectural plan, design, and height of the building to be 29 stories above and eight below ground (*Fig. 7.2*). Because of the long-term stop of the construction and change of the architectural design, the architectural drawings, construction documents, and site inspection reports had to be reviewed. The owner hired a structural engineering company to conduct an extensive field investigation, repair, and

Fig. 7.1: Site view when purchasing the building

Fig. 7.2: Perspective and real view of the rehabilitation building

strengthening program. In 2006, large-scale repair and strengthening work was carried out following the results of the field investigation, document review, and structural analysis.

7.2 Description of the structural rehabilitation

The original building had a structural system with steel-reinforced concrete composite system above the ground and reinforced concrete rigid frame system below the ground. There were two separate core structures, containing elevators, stairs, and service shafts, located at the side of the building. A schematic plan of the original building is shown in *Fig. 7.3*.

After purchasing the building, the group needed to change the architectural plan and design of the building for more efficient use. They changed the architectural plan and basic structural system as shown in *Fig. 7.4*. Because the constructions under the first floor to B7th floor were already finished, design change was mainly focused on the structure above the ground. Newly

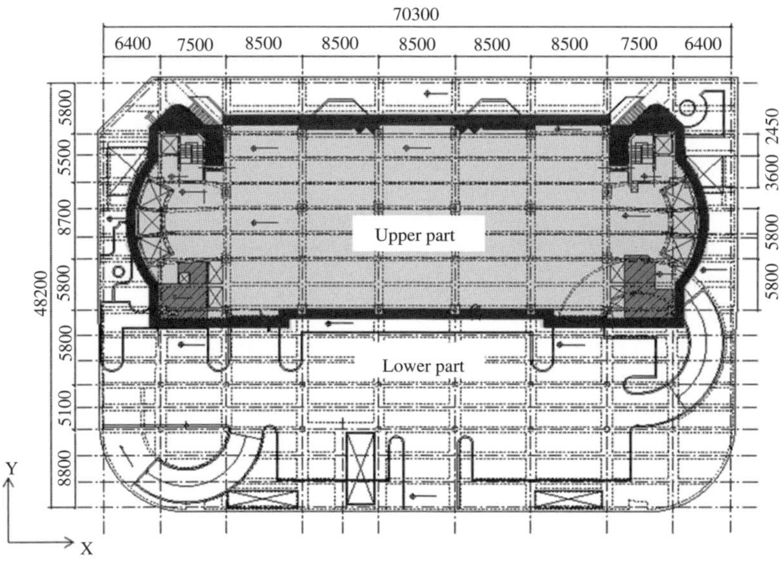

Upper part : 20 stories, side core *Lower part : 1 story, Entrance*

Fig. 7.3: Plan design before rehabilitation (Units: mm)

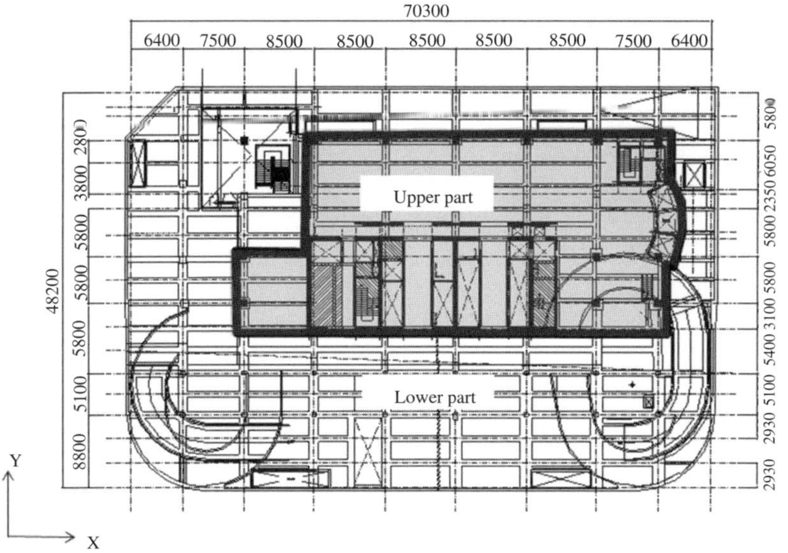

Upper part : 29 stories, center core *Lower part : 1 story, Entrance*

Fig. 7.4: Plan design after rehabilitation (Units: mm)

Floor	Original use	After rehabilitation	Summary of rehabilitation works
28th–29th	—	Pent house	Adding new structural story
1st–27th	Office	Office	Adding centre core First floor sunken garden
B1–B2	Office/parking	B1: Office B2: Parking	B1: adding transfer girder sunken garden B1–2: original stair demolition adding centre core
B3	Parking	Parking	Adding transfer girder Original stair demolition Adding centre core
B4–B7 mezzanine	Parking	Parking	Original stair demolition Adding centre stair
B7	Mechanic/ electric room	Mechanic/ electric room	Original stair demolition Adding centre stair

Table 7.1: Description of architectural story function and structural change

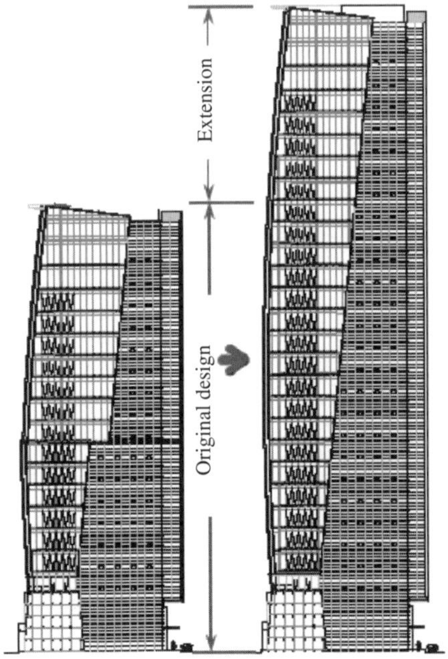

Fig. 7.5: Structural changing items

designed building story function and structural changing items are shown in *Table 7.1* and *Fig. 7.5.*

Following the change of the architectural plan and building function, the main structural rehabilitation work was roughly divided into two categories. One is the location of the core. Original location of the core was on each side of the building, but the new location of the core was changed to the centre of the building. The other is the change of the load conditions.

Increasing the story and changing the service loads caused insufficient nominal strength of the original members. Following these structural needs, the original finished construction region, that is, the first floor to B7th floor required newly installed transfer girders for the support of the centre core and strengthening of the beams, slabs, and columns.

Summary of structural changes is shown in *Table 7.2.*

7.3 Results of the field investigation

The building was constructed up to first floor slab, and construction was stopped for 13 years from 1993 to 2006. As a beginning of the rehabilitation project, existing conditions were investigated for check-up of the building's health.[4] The project was off to a good start because

Description		Before rehabilitation	After rehabilitation
Location		First Street ShinmunRo, ChongroGu, Seoul, Korea	
Site area (m^2)		3959	3959
Construction area (m^2)		1361	1361
Total building area (m^2)		60 628	60 628
Structural system	Underground	RC	RC
	Above ground	SRC	SRC
Scale	Underground	Seven floors	Eight floors
	Above ground	20 floors	29 floors
	Height (m)	About 80	118.8
Core	Location	Dual core	Centre core
	Anchorage	Foundation	Transfer at B3 floor
Structural rehabilitation summary	Space use change	First floor—demolition and reconstruction Underground—strengthening	
	New core	Demolition of existing member and construction of transfer girder (B1–B3 floor)	
	Foundation	Increasing design bearing force ($f_e = 3.75$ MPa) Demolition of existing floating anchor	

Table 7.2: Summary of structural changes

most structural drawings, including the design loads, material properties, member layout, and construction details, were available. Hence, the purpose of field investigation was to check the member layout, size, and material properties and to determine the extent of damage by visual inspection and non-destructive testing methods. Increasing member design load as a result of changing the architectural plan and story was another problem, so the field investigation was limited to the safety condition following the existing drawings.

After the document review and field investigation, it was concluded that the existing building structure was capable of serving in its original capacity. However, the members were partially deteriorating as time progressed.

Most symptoms on the structural deterioration were not severe and suspected to the material's property not the structural problems. As shown in *Fig. 7.6*, some visible cracks were found

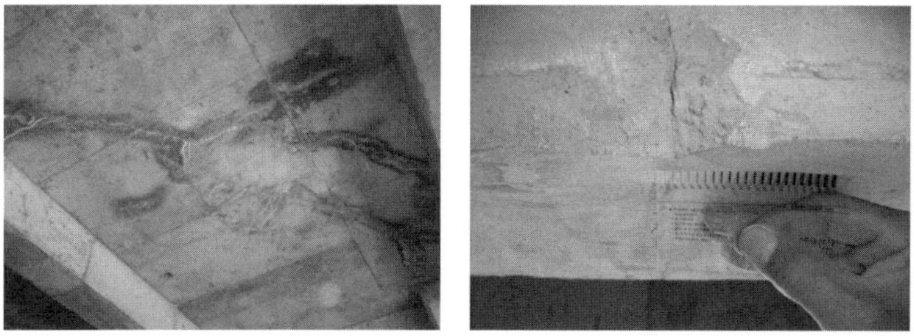

Fig. 7.6: Cracks at the slab and beam in the existing structure

in the slabs and beams. Because the patterns of the cracks were not typical and the size was not large, it was concluded that only the execution of crack repair was sufficient for the rehabilitation project. Some parts of the structure were spalling, delaminating, and containing honeycombs as shown in *Figs. 7.7 and 7.8*. These may have been caused by shallow cover and construction faults. *Figure 7.9* shows the overall site view of the existing structure before rehabilitation.

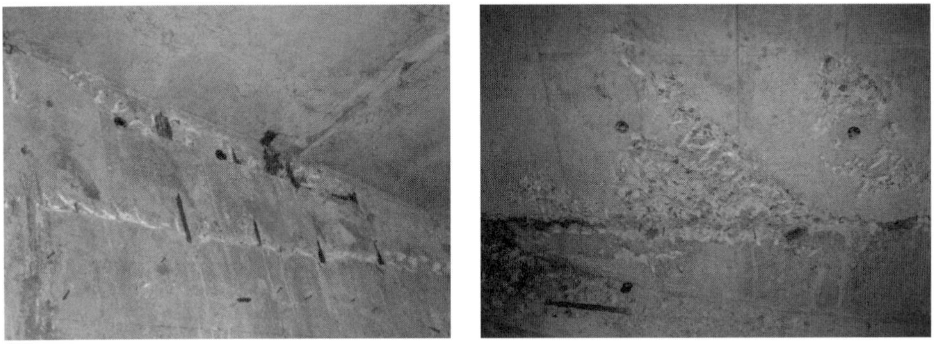

Fig. 7.7: Spalling and delamination in the existing structure

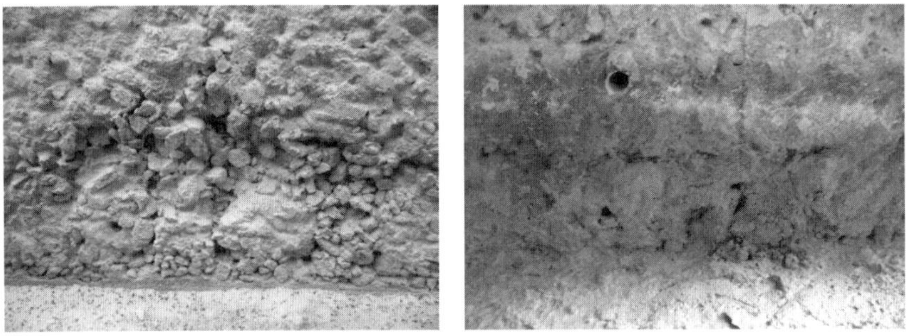

Fig. 7.8: Honeycomb in the existing structure

Fig. 7.9: Site view before rehabilitation

7.4 Structural analysis before and after repair, design of sections, and codes

7.4.1 Design codes

After the field investigation was almost completed, the structural engineers evaluated the load-carrying capacities of the existing structural members, and they proposed a rehabilitation program for modifications and upgrades to make the building suitable for the new architectural plan and design.

The original structural system was designed by KCI 1988 code (Korea Concrete Institute—Building Code Requirements for Reinforced Concrete Structures), and constructed up to 1993 following the original structural drawings. KCI 1988 code was basically based on the ultimate strength design method. [1]

The design code for structural rehabilitation was mainly based on KBC-S 2005 (Korean Building Code—Structure 2005) and KCI 2003 code (Korea Concrete Institute—Revision of the KCI 1988 Code) which were very similar to ACI 318-99 code.[2], [3] Review of building vibration was based on ISO 2631/1.

Structural analysis and member design were carried out using the MIDAS program, which was a structural analysis program developed by a Korean company.

7.4.2 Structural analysis results before repair

The results of structural analysis indicated that the foundation and underground exterior wall were mostly safe regardless of the rehabilitation. However, load-carrying capacities of some slabs and beams were insufficient because of structural layout and design code changes. Some columns under the main structure also had insufficient bearing capacities because of the change in number of stories from 20 to 29.

Concrete compressive strength was 27 MPa, yield strength of reinforcing steel was 400 MPa (SD40), and structural steel was SM490 and SS400 ($f_y = 330$ MPa, $f_y = 240$ MPa each). The foundation was drop type mat foundation, and rock anchor system was used for prevention of floatage.

Following the analysis results, insufficient bearing capacity member lists are summarized in *Tables 7.3 and 7.4.*

7.4.3 Structural analysis results after repair

For the resistance of lateral loads after rehabilitation, several structural systems were investigated. A central concrete core system was finally selected because of its effectiveness of lateral displacement control and architectural space program.

Following the structural analysis results, the possibility of core location change from each side of the building to the centre was verified and re-designed, and some individual members were also strengthened to resist the new architectural design (*Tables 7.3 and 7.4, and Fig. 7.10*).[5, 6].

Member list	Floor	Section (mm)	Required strength			Nominal strength			Section after strengthening (mm)
			P_u (MN)	M_{ux} (KNm)	M_{uy} (KNm)	ΦP_n (MN)	ΦM_{nx} (KNm)	ΦM_{ny} (KNm)	
C_2	−2	1000 × 1200	21.73	190	110	18.06	190	110	1200 × 1400
	−3 to −4	1000 × 1300	24.63	270	160	21.86	290	170	1300 × 1400
	−5 to −6	1000 × 1400	27.50	240	140	23.69	250	180	1400 × 1400
	−7	1000 × 1500	29.16	210	10	24.97	220	10	1500 × 1400
C_3	−1	900 × 900	24.85	120	0	11.60	70	0	1300 × 1300
	−2	900 × 900	25.79	270	100	12.53	160	60	1300 × 1300
	−3	900 × 900	27.32	340	130	14.48	220	80	1300 × 1300
	−4 to −5	900 × 900	31.39	360	30	14.48	200	20	1300 × 1300
	−6 to −7	950 × 950	31.15	330	70	15.94	210	40	1350 × 1350
C_{4A}	−1	900 × 900	15.62	250	100	11.29	180	70	1300 × 1300
	−2	900 × 900	16.85	30	20	12.53	30	20	1300 × 1300
	−3	900 × 900	18.24	50	210	14.48	50	200	1300 × 1300
	−4 to −5	900 × 900	20.98	40	250	14.48	30	210	1300 × 1300
	−6 to −7	950 × 950	26.91	70	270	15.94	50	200	1350 × 1350
C_{7C}	−1 to −2	500 × 500	4.84	30	80	3.59	150	80	900 × 500
	−3 to −4	650 × 650	8.00	10	420	6.71	10	370	1050 × 650
	−5 to −6	750 × 750	11.17	0	0	8.51	0	0	1150 × 750
	−7	800 × 800	12.78	0	10	7.74	890	20	1200 × 800
C_{9A}	−2	1400 × 1000	25.79	70	220	20.53	70	220	1400 × 1400
	−3	1500 × 1000	28.73	140	230	24.43	150	240	1500 × 1400
	−4	1500 × 1000	28.73	140	230	24.43	150	240	1500 × 1400
	−5	1600 × 1000	31.39	50	200	26.26	50	210	1600 × 1400
	−6	1600 × 1000	31.39	50	200	26.26	50	210	1600 × 1400
	−7	1700 × 1000	33.43	50	200	25.70	50	190	1700 × 1400
C_{10A}	−2 to −4	800 × 800	19.42	100	160	9.51	60	90	1000 × 1000
	−5	800 × 800	19.42	100	160	9.75	60	100	1000 × 1000
	−6 to −7	800 × 800	19.42	100	160	9.94	60	100	1000 × 1000

Table 7.3: Column list summary of insufficient bearing capacity

Member list	Moment ratio*			Shear ratio*			Grade**	Insufficient stress	Strengthening
	Ext.	Cent.	Int.	Ext.	Cent.	Int.			
1G3 [6/A–B]	0.62	0.94	1.13	0.72	0.38	1.36	E	Moment/shear	Steel plate
1G3 [7,8,9/A–B]	0.37	0.91	1.00	0.68	0.30	1.21	D	Shear	Steel plate
1G4 [6,7/B–C]	1.39	0.12	0.16	1.17	0.94	0.27	D	Moment/shear	Steel plate
1G15 [2/C–D]	0.72	0.53	0.95	1.00	0.16	1.21	D	Shear	Steel plate
B1G3 [4,5,6/A–B]	0.69	1.01	1.26	0.62	0.33	1.12	D	Moment/shear	Steel plate
B2∼3G3 [6/A–B]	1.69	1.19	0.00	1.38	1.20	0.70	E	Moment/shear	Steel plate
B4∼5G3 [6/A–B]	1.74	1.42	0.00	1.39	1.21	0.69	E	Moment/shear	Steel plate
B6G3 [7/A–B]	1.27	1.01	0.00	1.04	0.86	0.47	E	Moment	Steel plate

*Moment and shear ratio means the ratio of nominal strength divided by the design strength.

**Grade means the safety degree on the *Korean Guide Manual of Safety Inspection and Diagnosis*.[4] It consists of five grades. "A" means the best condition, and "E" means the worst condition.

Table 7.4: Beam list summary of insufficient bearing capacity

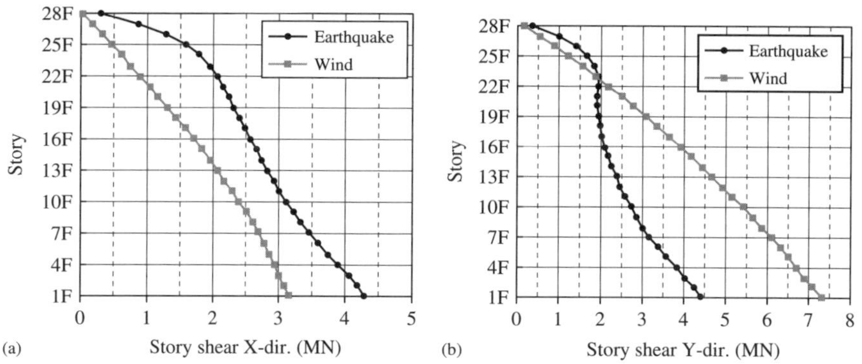

Fig. 7.10: Analysis results of story shear after rehabilitation

7.5 Repair strategies

As mentioned before, construction of this building stopped for 13 years. Several deterioration symptoms were found on the existing structural members by field investigation. However, the structural engineer decided that the extent of symptoms were not severe and not affecting structural safety after investigating the field inspection results. Hence, repair works for increasing the durability were achieved.

Main repair work was classified by the deterioration symptoms of the structural members.[6] As shown in *Table 7.5*, cracks in the structural members were repaired by surface caulking or epoxy injection methods according to the maximum width of the crack. Surface caulking was executed by applying the epoxy putty on the surface of the members without any injection

Symptoms		Repair methods	Materials
Crack	Minor cracks on non-bearing structure	No repair	
	Minor cracks whose maximum width is lower than 0.2 mm	Surface caulking	Epoxy putty
	Heavy cracks whose maximum width is larger than 0.3 mm	Epoxy injection	Epoxy putty, Epoxy for injection
Concrete spalling/ delamination/ honeycomb	Exposed re-bar	Patching method	Protective coating agent, Bonding agent, Polymer mortar
	Unexposed re-bar		Bonding agent, Polymer mortar

Table 7.5: Repair method and materials for symptoms

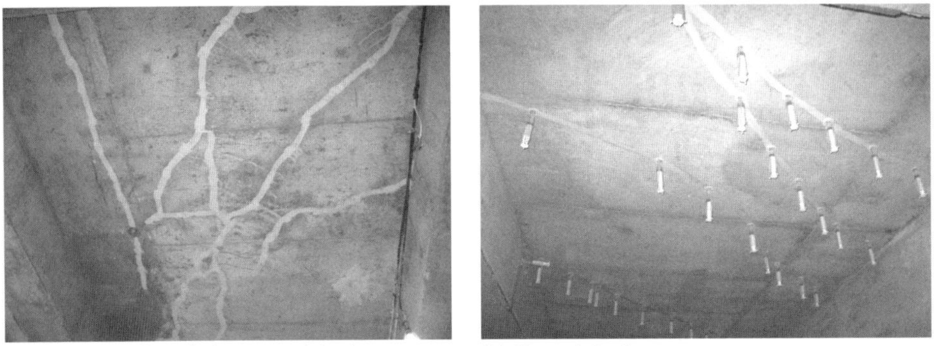

(a) Surface caulking before epoxy injection *(b) Epoxy injection*

Fig. 7.11: Methods of crack repair

Fig. 7.12: Patching method at the honeycomb and concrete spalling region

materials. The structural engineer decided that this work is sufficient to increase the durability of the structure in the case of minor crack width.

Concrete sections damaged by spalling, delamination, and honeycomb were repaired by patching method. The patching method calls for removing all deteriorated concrete, exposing and cleaning rusted steel reinforcement, coating the bars with a corrosion-protective coating agent, and patching the areas of removed concrete by polymer mortar.

The view of repair work is shown in *Figs. 7.11 and 7.12*. *Figure 7.11* shows the methods of crack repair at the bottom of slab by surface caulking and epoxy injection. *Figure 7.12* shows the view of repaired members after repair by patching method.

7.6 Strengthening, rehabilitation strategies, and detailing

For the renewal of the structure, demolition of the existing first floor slab for the new architectural design, and demolition and adding the bearing walls for the expansion of ramp for the vehicles were required.

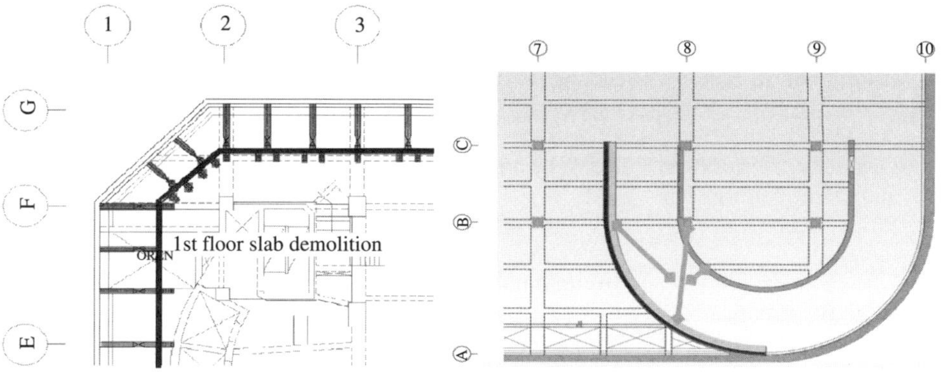

(a) B1[st] wall temporary shoring *(b) B2[nd] ramp temporary shoring*

Fig. 7.13: Temporary shoring plan

Fig. 7.14: Temporary shoring

To achieve this structural rehabilitation work, temporary shoring was required for the support of exterior wall to resist soil pressure during the construction period. The ramp bearing wall also required temporary support for resisting existing loads. The structural engineer suggested temporary supporting system using steel members. Schematic drawings of temporary support are shown in *Fig. 7.13*, and view of temporary support is shown in *Fig. 7.14*.

Following the structural analysis, some members were required to reinforce for load-carrying capacity. *Table 7.6* summarized the strengthening methods by members. Because of the change in the main building location and adding floors, many columns needed an increase for bearing capacity. Location of the columns requiring strengthening is shown in *Fig. 7.15*. The structural engineers chose section enlargement method in this project because this was the most common method in Korea, and they could easily obtain the required quality. Other benefits of this method were that it could preserve the existing column, get a reduction in slenderness, and not disturb the rehabilitation design. *Figure 7.16* shows a detail of exterior and interior column enlargements, and anchorage system to the foundation. In all cases, composites action between the new and the existing concrete was achieved by a combination of bond and anchoring dowels. *Figure 7.17* shows the sequences of section enlargement method used in this project, and *Figures 7.18 and 7.19* represent the joint detail between existing and adding new members.

Member	Causes	Strengthening concept	Strengthening method
Slab	Large span slab	Shortening the span	Adding steel beams
	Insufficient re-bar	Strengthening moment capacity	Adding bonded FRP (Fiber Reinforced Polymer)
	Concrete deterioration and insufficient re-bar	Demolition and adding new slab	Demolition and adding new slab
Beam	Positive moment increase	Flexural strengthening under the soffit and shear strengthening	Adding bonded steel plate
	Negative moment increase	Flexural strengthening above the beam	Adding embedded steel plate
Column	Load increase	Increasing the capacity	Column section enlargement

Table 7.6: Strengthening methods classified by member

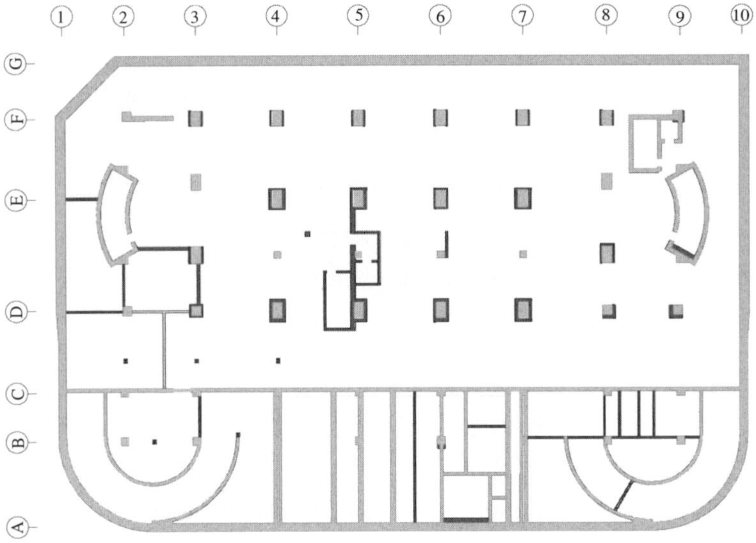

bold line : adding or enlarging section

Fig. 7.15: Structural plan (B7th)

Some beams and girders required flexural and shear strengthening because of changing load conditions. Many techniques for upgrading beams were applicable. On the basis of successful prior practices, the solution they suggested was using steel plates from 6 to 12 mm thick. Adding bonded steel plates could be done quickly, with minimum disturbance of the existing construction. It is also easy to visualize things that could go wrong with epoxy application. To achieve strengthening quality, this system must keep the demands as follows; at first, the surface must be thoroughly prepared, with mill scale and contaminants removed and the surface roughened. Second, the epoxy should have a bond strength equal to or exceeding that

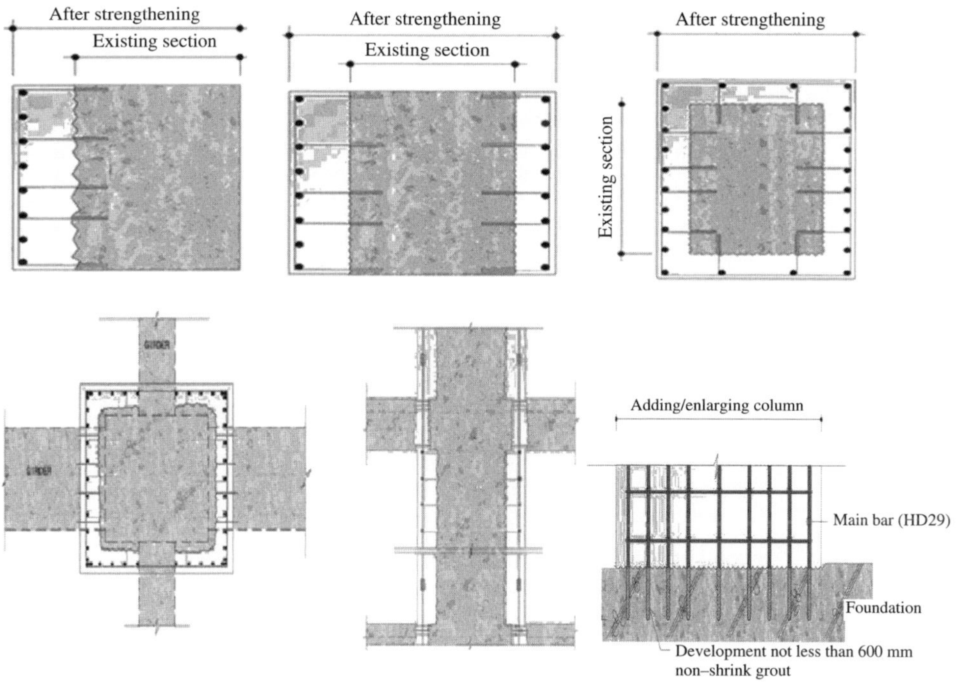

Fig. 7.16: Strengthening concrete columns by section enlargement

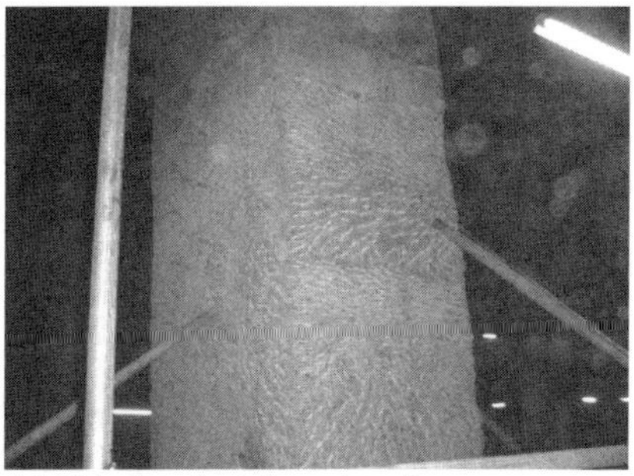

(a) Chipping of existing column face

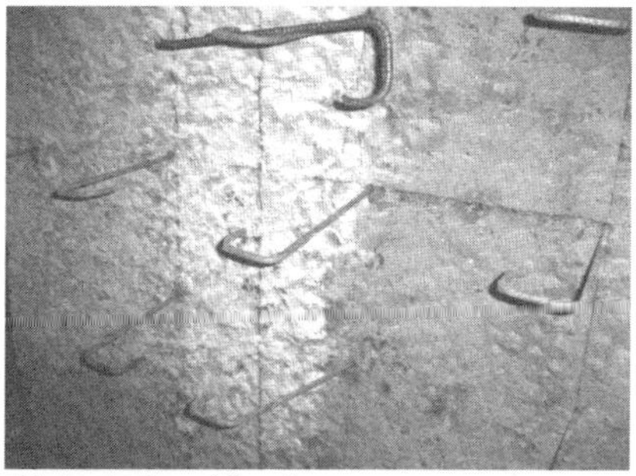

(b) Anchoring of dowel

(c) Apply additional reinforcement

(d) Casting concrete

Fig. 7.17: Steps for section enlargement of column

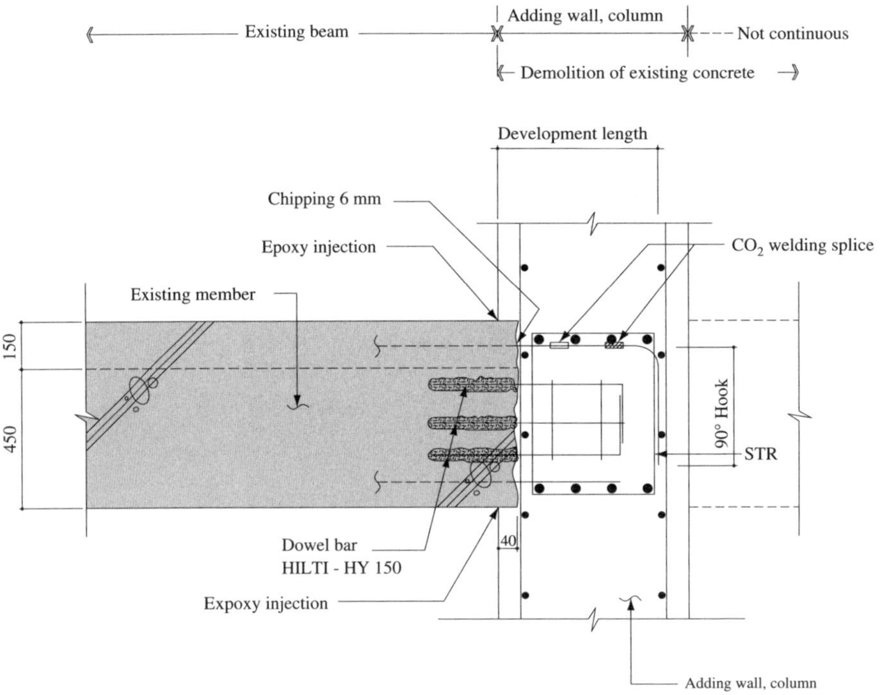

Fig. 7.18: Joint detail between existing beam and adding column (Units: mm)

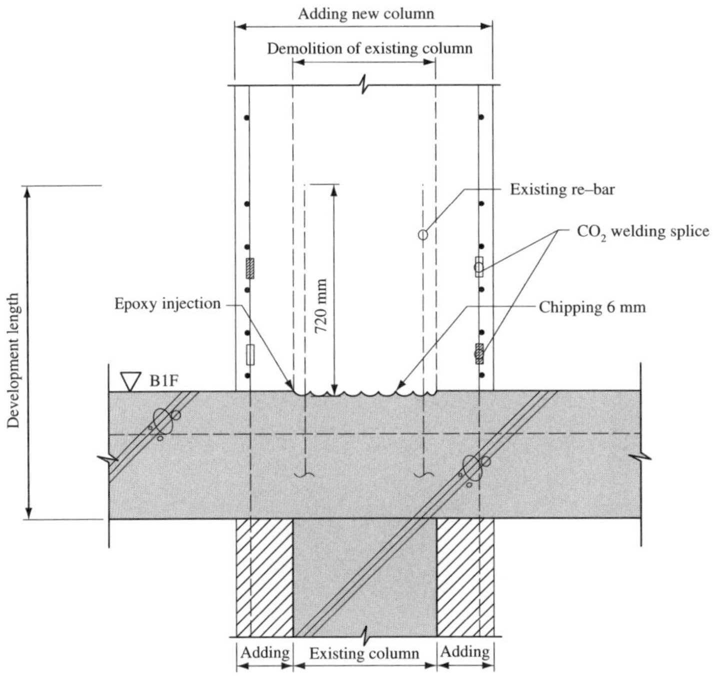

Fig. 7.19: Joint detail of existing column demolition and adding new column

of concrete. Finally, the reinforcing steel plate should be long and thin enough to avoid brittle plate separation from the existing concrete.

The location that required upgrading of beams is shown in *Fig. 7.20*. Some members needed flexural and shear strengthening, and some required flexural strengthening only following the load-carrying capacity and analysis results.

The view of upgrading beams under the soffit and shear strengthening is shown in *Fig. 7.21*. Some girders needed to upgrade the negative moment capacity. The structural engineer suggested using the embedded steel plate method. The sequence of construction was very similar to the bonded steel plate method. However, the bonded steel plate was not located

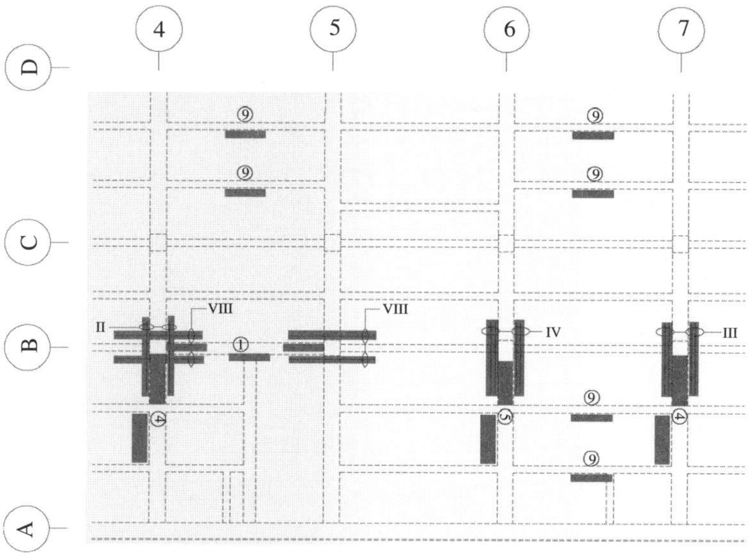

①..⑨. bonded steel plate for flexural and shear strengthening
I-VII: adding embedded steel plate in the negative moment region

Fig. 7.20: B6th floor structural plan

Fig. 7.21: Adding bonded steel plate in beams

(a) Steel plate embedment above the beam *(b) Epoxy mortar casting*

Fig. 7.22: Adding embedded steel plate

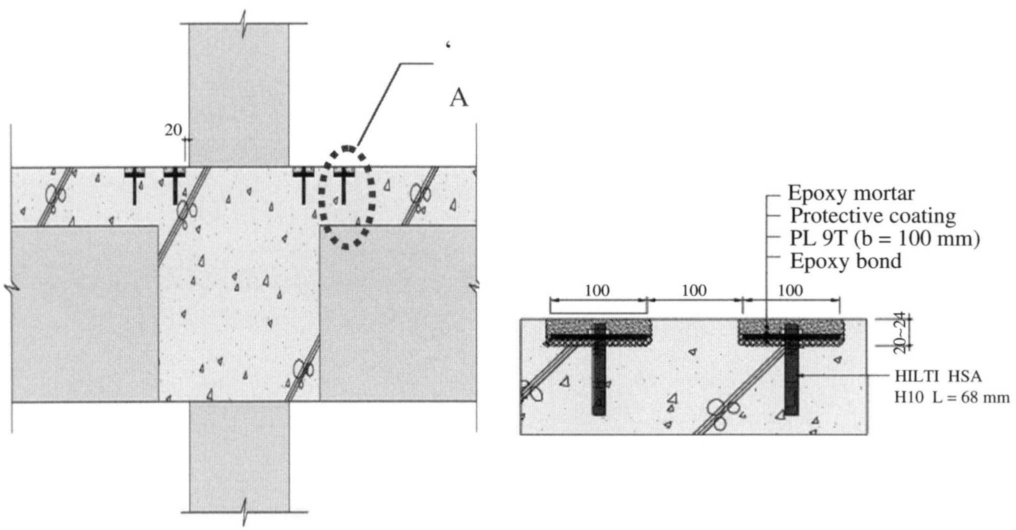

(a) Section *(b) Detail 'A'*

Fig. 7.23: Detail of adding embedded steel plate

under the member but above, and filled with epoxy mortar on the plate. *Figures 7.22 and 7.23* show the embedded steel plate method.

Removing the existing side core and adding the new centre core following the changed architectural design were large-scale rehabilitation works in the already constructed region. The core was originally located in the side of the building, but the architects newly designed the centre core for the changed architectural function. Before constructing the new centre core, demolition work of the existing side core was carried out as shown in *Fig. 7.24*.

(a) Demolition of side core at B4 floor (b) Demolition view at 1st floor

Fig. 7.24: Demolition of existing underground core

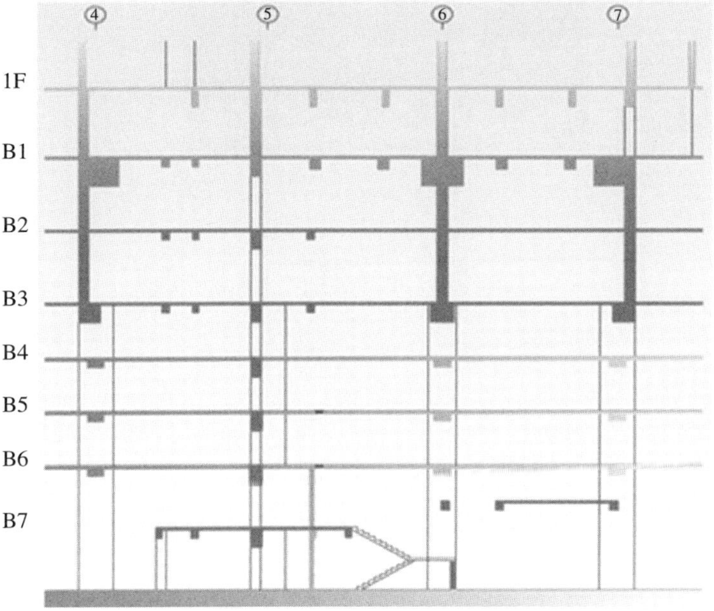

Fig. 7.25: Section of centre core

For the centre core construction, it is required to remove the members in the new core construction area. After the removal of the slabs and beams, a new transfer girder was added from B1st to B3rd floor as shown in *Fig. 7.25*. New centre stair was added from B4th to B7th floor, and a mezzanine floor was added in the B7th floor. The design of form, shoring, and adding members was based on the results of structural analysis.

The added members were connected to the adjoining frame by drilled-in dowels and welding of reinforcements. The step of adding the new centre core is roughly shown in *Fig. 7.26*, and joint detail of adding transfer girder is shown in *Fig. 7.27*.

(a) Installation of reinforcement
(b) Formwork
(c) Casting concrete
(d) Completion

Fig. 7.26: Adding new centre core

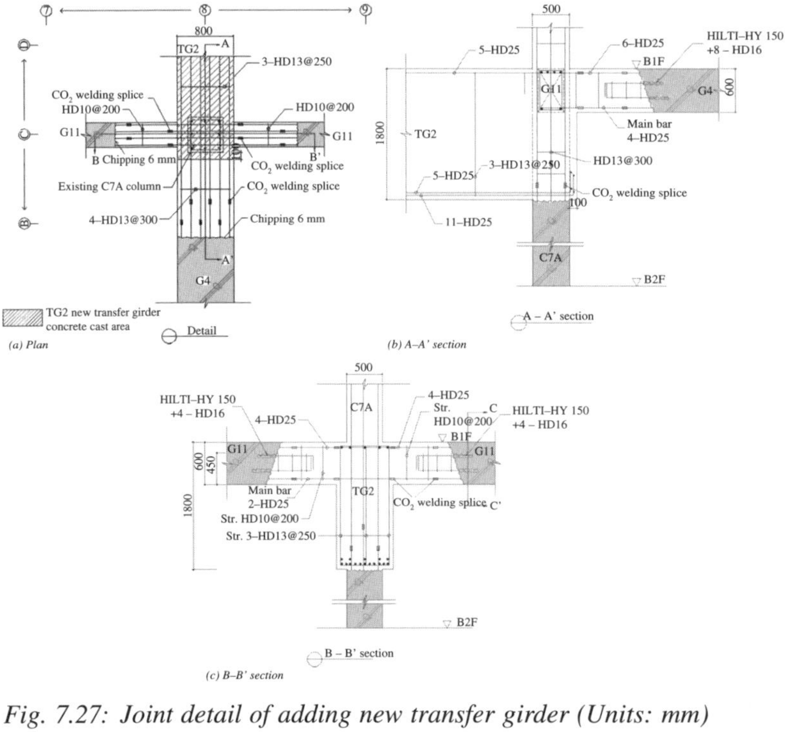

Fig. 7.27: Joint detail of adding new transfer girder (Units: mm)

7.7 Summary

Rehabilitation usually means an upgrading required to meet the present needs. Functionally inadequate buildings are common candidates for rehabilitation. In this project, the new owner purchased a partly built building for which construction was not finished. Their requirements for headquarters were different from the original building design functions. They hired an architect and a structural engineer and a revision of the architectural function was commissioned. They investigated the original drawings and the new owner's requirements. Finally, they suggested the revised architectural design and rehabilitation methods after several meetings between the architects and structural engineers. The selected scheme of strengthening below the ground for this project evolved from many practices, studies, and concepts.

The project was bid in early 2006, and rehabilitation for the existing structure was successfully completed on schedule at the end of 2006. Before repair and strengthening, this building's safety grade on the *Korean Guide Manual of Safety Inspection and Diagnosis* was "D" when considering the change of architectural plan and load conditions.[4] Grade "D" indicates a condition that many structural members have insufficient strength capacity and require repair and strengthening. After the rehabilitation works, safety grade of this building improved to grade "B". Grade "B" indicates a good condition that partial non-structural members require continuous inspection for maintenance.

7.8 Acknowledgements

We hereby appreciate the co-operation of Kumho group, field manager Jae-Ung Jung, and structural engineer In-Sik Im for offering several rehabilitation data.

References

[1] Korea Concrete Institute. *Structural Design Code and Commentary for Reinforced Concrete*. Korea Concrete Institute, Seoul, 1988.

[2] Korea Concrete Institute. *Structural Design Code and Commentary for Reinforced Concrete*. Korea Concrete Institute, Seoul, 2003.

[3] AIK. *Korean Building Code and Commentary*. Architectural Institute of Korea, Seoul, 2005.

[4] KISTEC. *Guide Manual of Safety Inspection & Diagnosis*. Korea Infrastructure Safety and Technology Corporation, Seoul, 1998.

[5] Teng J.G., Chen J.F., Smith S.T., Lam L. *FRP Strengthened RC Structures*. John Wiley and Sons, England, 2001.

[6] Newman A. *Structural Renovation of Buildings Methods, Details, and Design Examples*. McGraw-Hill, New York, USA, 2001.

Chapter

8

Strengthening the Murhasaari Bridge with External Prestressing, Finland

Ilkka Vilonen, L.Sc., Ramboll Finland Ltd, Tampere, Finland

Abstract: The Murhasaari Bridge is located on Highway 11 between Pori and Tampere, which is one of the most heavy traffic roads in the Finnish road network, carrying total loads exceeding 140 t. The bridge crosses a lake near Nokia with three spans 26 + 52 + 26 m. The deck width is 10.5 m from railing to railing.

The bridge piers have direct foundations on solid moraine strata, and there are no signs of movements in the substructure. The bridge superstructure is a continuous box girder, made of reinforced concrete without prestressing. The cross section has three cells in the side spans, changing to two cells in the main span.

The bridge was constructed in 1962, when the design traffic load was only 140 kN axle load and distributed load 4 kN/m^2. These loads are much smaller than the loads used today.

Soon after construction, routine inspection detected deflections of the midspan, which also impaired proper functioning of the expansion joints at the abutments. Repair in 1977 included replacement of the moisture isolation and asphalt on the bridge deck, injection of some cracks with epoxy, and repair of expansion joints and edge beams.

Between 1976 and 1994, the midspan deflection increased by 110 mm, and in the following years, special inspection and load testing disclosed extensive cracking and insufficient load-carrying capacity.

The bridge was strengthened in 1999 with post-tensioning using external cables, placed inside the box chambers. At both ends of the deck, new cross beams were constructed to anchor the post-tensioning cables. The original cross beams along the bridge were used as deviators, to get suitable bending moments and shear forces opposite to the original ones. As a result, the midspan deflection was reduced by 50 mm, the growth stopped, and the need for frequent repair of the expansion joints eliminated.

Strengthening the bridge with post-tensioned external cables was found to be a very effective way to increase the bearing capacity. Bridges requiring this kind of rehabilitation are usually those designed under previous design codes. Especially, this type of reinforced concrete box-girder bridges can be easily strengthened by mounting the tendons inside the boxes. This technique was found to be cost effective as well. The paper presents details about the bridge and external post-tensioning.

NOTE: This is the abstract. Full paper will be available in the electronic version of SED12.

Appendix

A

List of Articles from IABSE—SEI Journal Related to Topics of IABSE SED 12

This Appendix mainly includes a list of articles published in IABSE SEI Journal, which are related to topics of SED 12 (rehabilitation, repair, retrofit, strengthening, upgrading, . . . of structures). The listed articles could present additional case studies to those presented in SED 12. The appendix also includes information on where to order other IABSE publications related to the topics of SED 12 (e.g. IABSE Conference Proceedings, and other SEDs). Additional case studies relevant of the topics of SED 12 could be found also in the followings:

(I) Articles from IABSE SEI Journal: from 1991 to 2009. To obtain a copy of the articles refer to the website of IABSE SEI: http://www.ingentaconnect.com/content/iabse/sei

(II) IABSE Conferences: http://www.iabse.org/publications/iabsereports/index.php

(III) IABSE SED: http://www.iabse.org/publications/onlineshop/index.php

(IV) IABSE E-Learning: (Audio Visual presentations), Refer to: http://www.iabse.org/

(I) Articles from IABSE SEI Journal

The articles from IABSE SEI Journal on topics relevant to SED 12 are organized in groups as follows. The groups/categories should be considered as tentative. There are no definite limits or rigid boundaries between the activities in the groups, and there is lots of interaction. The groups are not necessarily organized in chronological order.

Note: In the following, the numbers in parentheses(1)–(41) refer to the relevant parts in the list of articles. For definitions (partial list), please refer to "*Introduction*", section on *Terminology*, and to Appendix B (section X).

Group (A): Maintenance(1) (Policy, Strategy, Optimization, Cost), Management systems(2) (Bridges, Building, Facilities), Operation(3). This group is related mainly to the policies and planning of the activities in the post-construction phase, aiming at maintaining and/or improving structural performance.

Group (B): Inspection(4), Monitoring(5), Testing(6) and Load Tests, Non-destructive Testing. This group is related mainly to observations and investigations carried out on the structures.

Group (C): Assessment(7), Evaluation(8), Extending Service Life(9), Load Capacity(10), Performance(11), Robustness(12), Safety(13). Additional relevant keywords also include: Analysis, Appraisal, Bridge Sufficiency, Durability, Investigation, Rating, ... This group is related mainly to condition assessment and evaluation of structural performance.

Group (D): Corrosion[14], Cracking[15], Damage[16], Deterioration[17], Fatigue[18], Fracture[19], Scour[20], Vibration[21], Weathering[22]. This group is related to deterioration in materials and/or structures, which may result in a reduction of structural performance. Relevant keywords also include: Abrasion, Bleeding, Deficiency, Degradation, Delamination, Discoloration, Distress, Erosion, Leak, Oxidation, Pounding, Rupture, Rusting, Scaling, Spalling, Surface Defects, Unseating, Wash Out, Wear, ...

Group (E): Prevention[23], Protection[24]. Additional relevant words include: Coating, Impregnation, Painting, Remove debris from joints, Resurfacing, Sealing (cracks deck surface), Touch up minor paint defects. The activities might include application of paints on surfaces, and/or minor changes in member dimensions. This group is related mainly to activities relevant to preventive maintenance.

Group (F): Preservation[25], Refurbishment[26], Rehabilitation[27], Remedial[28], Renew[29], Repair[30], Replacement of components[31], Restoration[32], Retrofit[33], Stiffening[34], Strengthening[35], Upgrading[36], Widening[37]. Relevant keywords also include: Conservation, Addition (of structural members, bracing for example), Increase Load Capacity, Modification, Reinforcement, Renovation. This group is related to the changes in the dimensions of structural members (large structural intervention) to restore and/or upgrade the structural performance.

Group (G): Rebuilding[38], Reconstruction[39], Replacement of Structures[40].

Group (H): Alteration, Change of Occupancy or Use, Reuse[41], ...

Group (A):

(1) Maintenance

1. Probability-Based Assessment and Optimised Maintenance Management of a Large Riveted Truss Railway Bridge, O'Connor, Alan; Pedersen, Claus; Gustavsson, Lars; Enevoldsen, Ib, SEI, Vol. 19, No. 4, Nov. 2009, pp. 375–382.
2. An Introduction: Operations, Maintenance and Repair of Structures, Sobrino, Juan A., SEI, Vol. 17, No. 4, Nov. 2007, pp. 328.
3. Performance-based Standard Specifications for Maintenance and Repair of Concrete Structures in Japan, Ueda, Tamon; Takewaka, Koji, SEI, Vol. 17, No.4, Nov. 2007, pp. 359–366.
4. Economic Efficiency of Modern Timber Bridges—Life Expectancy and Costs of Maintenance, Gerold, Matthias, SEI, Vol. 16, No. 3, Aug. 2006, pp. 261–267.
5. Why "waste" money on maintenance? Popovic, Predrag L., SEI, Vol. 14, No. 2, May 2004, pp. 71.
6. Lifetime Bridge Maintenance Strategies Based on System Reliability, Frangopol, Dan M.; Estes, Allen C., SEI, Vol. 7, No. 3, Aug. 1997, pp. 193–198.
7. Bridge Maintenance Needs and Costs in Sweden, Ingvarsson, Hans, SEI, Vol. 2, No. 3, Aug. 1992, pp. 202–205.

(2) Management Systems

1. Probability-Based Assessment and Optimised Maintenance Management of a Large Riveted Truss Railway Bridge, O'Connor, Alan; Pedersen, Claus; Gustavsson, Lars; Enevoldsen, Ib, SEI, Vol. 19, No. 4, Nov. 2009, pp. 375–382.
2. Bridge Management Systems, Hajdin, Rade, SEI, Vol. 8, No. 3, Aug. 1998, pp. 211.
3. Bridge Management for New York City, Yanev, Bojidar S., SEI, Vol. 8, No. 3, Aug. 1998, pp. 211–215.
4. Creating a Bridge Management System, Lauridsen, Jørn; Bjerrum, John; Andersen, Niels Hutzen; Lassen, Bjørn, SEI, Vol. 8, No. 3, Aug. 1998, pp. 216–220.
5. Condition Data and Bridge Management Systems, Hearn, George, SEI, Vol. 8, No. 3, Aug. 1998, pp. 221–225.

6. New Developments in Bridge Management Methodology, Das, Parag C., SEI, Vol. 8, No. 4, 1 Nov. 1998, pp. 299–302.
7. The Pontis Bridge Management System, Thompson, Paul D.; Small, Edgar P.; Johnson, Michael; Marshall, Allen R., SEI, Vol. 8, No. 4, 1 Nov. 1998, pp. 303–308
8. The BRIDGIT Bridge Management System, Hawk, Hugh; Small, Edgar P., SEI, Vol. 8, No. 4, 1 Nov. 1998, pp. 309–314.
9. The Finnish Bridge Management System, Söderqvist, Marja-Kaarina; Veijola, Magnus, SEI, Vol. 8, No. 4, 1 Nov. 1998, pp. 315–319.

(3) Operation

1. An Introduction: Operation, Maintenance and Repair of Structures, Sobrino, Juan A., SEI, Vol. 17, No. 4, Nov. 2007, pp. 328.

Group (B):

(4) Inspection

1. Risk-Based Inspection: An Introduction, Faber, Michael H., SEI, Vol. 12, No. 3, Aug. 2002, pp. 186.
2. Risk-Based Inspection: The Framework, Faber, Michael H., SEI, Vol. 12, No. 3, Aug. 2002, pp. 186–195.
3. Codified Risk-Based Inspection Planning, Soerensen, John D.; Faber, Michael H., SEI, Vol. 12, No. 3, Aug. 2002, pp. 195–199.
4. Risk-Based Inspection Planning of Offshore Installations, Goyet, Jean; Straub, Daniel; Faber, Michael H., SEI, Vol. 12, No. 3, Aug. 2002, pp. 200–208.
5. Optimum Lifetime Planning of Bridge Inspection and Repair Programs, Frangopol, Dan M.; Estes, Allen C., SEI, Vol. 9, No. 3, Aug. 1999, pp. 219–223.
6. Main Cable Strand Inspection of the George Washington Bridge, Trivedi, Arvind; Lynch, Marjorie, SEI, Vol. 6, No. 1, Feb. 1996, pp. 37–40.
7. Bridge Inspection Vehicles, Grüter, Rolf; Hoffmann, Rolf, SEI, Vol. 2, No. 2, May 1992, pp. 92–94.

(5) Monitoring

1. Condition Monitoring of Masonry Arch Bridges using Acoustic Emission Techniques, Tomor, Adrienn K.; Melbourne, Clive, SEI, Vol. 17, No. 2, May 2007, pp. 188–192
2. Monitoring the Akashi Kaikyo Bridge: First Experiences, Kashima, Satoshi; Yanaka, Yukikazu; Suzuki, Shuichi; Mori, Kunihisa, SEI, Vol. 11, No. 2, May 2001, pp. 120–123.
3. Monitoring of the Paraná River Bridge during Launching, Malite, Maximiliano; Takeya, Toshiaki; Gonçalves, Roberto Martins; de Sáles, José Jairo, SEI, Vol. 10, No. 3, Aug. 2000, pp. 193–196.
4. Design and Monitoring of the Flintshire Bridge, UK, Curran, Peter; Tilly, Graham, SEI, Vol. 9, No. 3, Aug. 1999, pp. 225–228.
5. Eigenfrequency Monitoring during Bridge Erection, Skrinar, Matjaz; Strukelj, Andrej, SEI, Vol. 6, No. 3, Aug. 1996, pp. 191–194.
6. Monitoring Medieval Structures in Pavia, Macchi, Giorgio, SEI, Vol. 3, No. 1, Feb. 1993, pp. 6–9.

(6) Testing

1. Model Test of Half-Through CFST Tied-Arch Bridge in the Process of Arch Rib Erection, Zhang, Ke-Bo; Zhang, Jian-Ren; Yan, Dong-Huang, SEI, Vol. 18, No. 4, Nov. 2008, pp. 396–402.
2. Load Tests of a Cable-Stayed Bridge in Coimbra, Portugal, Santos, Luís Oliveira; Min, Xu, SEI, Vol. 17, No. 4, Nov. 2007, pp. 337–341.

3. Fatigue Tests for a Riveted Steel Railway Bridge in Salzburg, Matar, Ehab Boghdadi; Greiner, Richard, SEI, Vol. 16, No. 3, Aug. 2006, pp. 252–260.
4. Dynamic Testing and Modeling of a 30-years' old Cable-Stayed Bridge, Gentile, Carmelo; Gennari-Santori, Alberto, SEI, Vol. 16, No. 1, Feb. 2006, pp. 39–43.
5. Concrete Cracking in Composite Bridges: Tests, Models and Design Proposals, Navarro, Miguel Gómez; Lebet, Jean-Paul, SEI, Vol. 11, No. 3, Aug. 2001, pp. 184–190.
6. Wind-Tunnel Tests of a Bridge Model with Active Vibration Control, Hansen, Henriette I.; Thoft-Christensen, Palle; Mendes, Pedro A.; Branco, Fernando A., SEI, Vol. 10, No. 4, 1 Nov. 2000, pp. 249–253.
7. Static and Dynamic Testing of Fibre-Reinforced Bridge Girders, Strehler, Bernhard; Künzle, Otto, SEI, Vol. 9, No. 4, 1 Nov. 1999, pp. 300–301.
8. In situ and Model Tests and Numerical Analysis of a Curved Cable-Stayed Bridge, Broquet, Claude; Brühwiler, Eugen, SEI, Vol. 9, No. 1, Feb. 1999, pp. 57–62.
9. Testing a Wood–Concrete Composite Bridge, Jutila, Aarne; Mäkipuro, Risto; Salokangas, Lauri, SEI, Vol. 7, No. 4, 1 Nov. 1997, pp. 275–277.
10. Dynamic Testing of Large-Scale Structures, Paultre, Patrick; Proulx, Jean, SEI, Vol. 7, No. 1, Feb. 1997, pp. 29–34.
11. Testing a Prototype Steel–Wood Composite Bridge, Bakht, Baidar; Krisciunas, Ray, SEI, Vol. 7, No. 1, Feb. 1997, pp. 35–41.
12. Advances in Ambient Vibration Testing: Ganter Bridge, Switzerland, Felber, Andreas; Cantieni, Reto, SEI, Vol. 6, No. 3, Aug. 1996, pp. 187–190.
13. Testing and Repair of Concrete Silos, Ajdukiewicz, Andrzej, SEI, Vol. 6, No.4, 1 Nov. 1996, pp. 278–281.
14. Testing of a 55 Year Old Bridge in Canada, Denis, Benoit; Mallikarjuna; Fafard, Mario; Beaulieu, Denis; Halchini, Constantin, SEI, Vol. 4, No. 1, Feb. 1994, pp. 29–33.
15. Segmental Aerial Structure Retrofit and Load Testing, Irshad, Mohammad; Poston, Randall W., SEI, Vol. 3, No. 2, May 1993, pp. 108–112.
16. Corrosion-resistant Fastenings in Road Tunnels—Field Tests, Böhni, Hans; Haselmair, Helmut; Übeleis, Andreas M., SEI, Vol. 2, No. 4, Nov. 1992, pp. 253–258.
17. Fatigue Tests of Monorail Guideway Girders, Okura, Ichiro, SEI, Vol. 1, No.2, May 1991, pp. 28–36.

Group (C):

(7) Assessment

1. Assessment Criteria for Existing Structures, Vrouwenvelder, Ton; Scholten, Nico, SEI, Vol. 20, No. 1, Feb. 2010, pp. 62–65.
2. Probability-Based Assessment and Optimised Maintenance Management of a Large Riveted Truss Railway Bridge, O'Connor, Alan; Pedersen, Claus; Gustavsson, Lars; Enevoldsen, Ib, SEI, Vol. 19, No. 4, Nov. 2009, pp. 375–382.
3. Assessment of the Seismic Resistance and Structural Safety of Existing Multistory Residential Buildings, Schwarz, Stephan; Leibovich, Edward; Yankelevsky, David Z., SEI, Vol. 19, No. 2, May 2009, pp. 165–173.
4. Fatigue Safety Assessment of Existing Steel Bridges in China, Wang, Chunsheng; Chen, Weizhen; Chen, Airong, SEI, Vol. 19, No. 2, May 2009, pp. 174–179.
5. Probability-Based Assessment of Existing Concrete Bridges—Stochastic Resistance Models and Applications, Wisniewski, Dawid; Casas, Joan Ramon; Henriques, Antonio Abel; Cruz, Paulo J.S., SEI, Vol. 19, No. 2, May 2009, pp. 203–210.
6. Reliability Assessment of Ageing Reinforced Concrete Structures—Current Situation and Future Challenges, Val, Dimitri V.; Stewart, Mark G., SEI, Vol. 19, No. 2, May 2009, pp. 211–219.
7. Assessment, Repair and Widening of the Villeneuve-Loubet Bridge, France, Vion, Philippe; Poineau, Daniel, SEI, Vol. 18, No. 4, Nov. 2008, pp. 343–345.

8. Civil Engineering Faculty of ETH Zurich, Switzerland—Structural Assessment for Earthquake, Zwicky, Daia; Bianchi, Carlo, SEI, Vol. 17, No. 4, Nov. 2007, pp. 342–351.
9. Preventive Medicine: Condition Assessment, Ratay, Robert T., SEI, Vol. 17, No. 2, May 2007, pp. 113–113.
10. Mechanism based Assessment of Masonry Arch Bridges, Kumar, Pardeep; Bhandari, N. M., SEI, Vol. 16, No. 3, Aug. 2006, pp. 226–234.
11. Design Oriented Approach for Progressive Collapse Assessment of Steel Framed Buildings, Vlassis, Anastasios G.; Izzuddin, Bassam A.; Elghazouli, Ahmed Y.; Nethercot, David A., SEI, Vol. 16, No. 2, May 2006, pp. 129–136.
12. Failure Assessment of Simply Supported Floor Slabs under Elevated Temperature, Omer, Echat; Izzuddin, Bassam A.; Elghazouli, Ahmed Y., SEI, Vol. 16, No. 2, May 2006, pp. 148–155.
13. Assessment of Damaged Post-Tensioning Tendons, Vill, Markus; Eichinger, Eva M.; Kollegger, Johann, SEI, Vol. 16, No. 1, Feb. 2006, pp. 44–48.
14. Pressure Vessel Explosions inside Buildings: Assessing Damage Using Stochastic Accident Simulation, Vaidogas, Egidijus R., SEI, Vol. 13, No. 4, Nov. 2003, pp. 249–253.
15. Assessment and Rehabilitation of Steel Bridges: An Introduction, Ghoshal, Amitabha, SEI, Vol. 12, No. 4, Nov. 2002, pp. 249–249.
16. Assessment and Measurement of the Forsmo Bridge, Sweden, Enevoldsen, Ib, Pedersen, Claus; Axhag, Frank; Johansson, Örjan; Töyrä, Björn, SEI, Vol. 12, No. 4, Nov. 2002, pp. 254–257.
17. Assessment of Old Steel Bridges, Germany, Geissler, Karsten, SEI, Vol. 12, No. 4, Nov. 2002, pp. 258–263.
18. Experience with Probabilistic-based Assessment of Bridges, Enevoldsen, Ib, SEI, Vol. 11, No. 4, Nov. 2001, pp. 251–260.
19. Dynamic Assessment of a Curved Cable-Stayed Bridge at the Malpensa 2000 Airport, Milan, Italy, Gentile, Carmelo; Martinez y Cabrera, Francesco, SEI, Vol. 11, No. 1, Feb. 2001, pp. 52–58.
20. Damage Assessment and Analysis of the 1997 Umbria-Marche Earthquakes, Spence, Robin; D'Ayala, Dina, SEI, Vol. 9, No. 3, Aug. 1999, pp. 229–233.
21. Impact-Echo Condition Assessment of Structures, Sansalone, M. J.; Streett, W. B., SEI, Vol. 6, No. 4, Nov. 1996, pp. 282–284.
22. Finite Element Assessment of Thin Shell Structures, Zahlten, Wolfhard; Gruber, arsten; Krätzig, Wilfried B., SEI, Vol. 4, No. 3, Aug. 1994, pp. 164–166.
23. European Laboratory for Structural Assessment, SEI, Vol. 3, No. 1, Feb. 1993, pp. 54–55.
24. Some Thoughts on the Reliability Assessment of Existing Structures, Schneider, Jörg, SEI, Vol. 2, No. 1, Feb. 1992, pp. 13–18.
25. Ethical and Social Aspects in Comprehensive Risk Assessment, Chakraborty, Aby:asachi; Gonen, Yitzhak G., SEI, Vol. 1, No. 1, Feb. 1991, pp. 38–41.

(8) Evaluation

1. Experimental Evaluation of Synchronisation in Footbridges due to Crowd Density, Araújo Jr, Moacir Carlos; Brito, Halane Maria Braga Fernandes; Pimentel, Roberto Leal, SEI, Vol.19, No. 3, Aug. 2009, pp. 298–303.
2. Reuse Dismantling and Performance Evaluation of Reusable Members, Fujita, Masanori; Iwata, Mamoru, SEI, Vol. 18, No. 3, Aug. 2008, pp. 230–237.
3. Evaluation of Fatigue Category of Riveted Steel Bridge Connections, Matar, Ehab B., SEI, Vol. 17, No. 1, Feb. 2007, pp. 72–78.
4. Load Capacity Evaluation of Existing Railway Bridges based on Robustness Quantification, Wisniewski, Dawid; Casas, Joan R.; Ghosn, Michel, SEI, Vol. 16, No. 2, May 2006, pp. 161–166.
5. Seismic Evaluation and Retrofit of Historical Churches, Sorace, Stefano; Terenzi, Gloria, SEI, Vol. 12, No. 4, Nov. 2002, pp. 283–288.

6. Evaluating the Central Piers of the Maracaibo Bridge, Venezuela, Sarcos Portillo, Antonio R.; García Legl, Hildrun F., SEI, Vol. 9, No. 2, May 1999, pp. 133–136.

(9) Extending Service Life

1. Extending the Life of Masonry and Concrete Arch Bridges, Sobrino, Juan A., SEI, Vol. 17, No. 4, Nov. 2007, pp. 328–336.
2. Economic Efficiency of Modern Timber Bridges—Life Expectancy and Costs of Maintenance, Gerold, Matthias, SEI, Vol. 16, No. 3, Aug. 2006 , pp. 261–267.
3. Life-Cycle Performance of Deteriorating Structures, Adey, Bryan; Brühwiler, Eugene; Frangopol, Dan M.; Faber, Michael H., SEI, Vol. 13, No. 3, Aug. 2003, pp. 202–204.
4. Extending Service Life through Pre-emptive Retrofits, Fisher, John W.; Connor, Robert J., SEI, Vol. 12, No. 4, Nov. 2002 , pp. 264–268.
5. Optimum Lifetime Planning of Bridge Inspection and Repair Programs, Frangopol, Dan M.; Estes, Allen C., SEI, Vol. 9, No. 3, Aug. 1999 , pp. 219–223.
6. Lifetime Bridge Maintenance Strategies Based on System Reliability, Frangopol, Dan M.; Estes, Allen C., SEI, Vol. 7, No. 3, Aug. 1997 , pp. 193–198.

(10) Load Capacity

1. Load Capacity Evaluation of Existing Railway Bridges based on Robustness Quantification, Wisniewski, Dawid; Casas, Joan R.; Ghosn, Michel, SEI, Vol. 16, No. 2, May 2006, pp. 161–166.
2. Load Bearing Capacity of Connections in Tempered Glass Structures, Bernard, Fabrice; Daudeville, Laurent; Gy, René, SEI, Vol. 14, No. 2, May 2004, pp. 107–110.

(11) Performance

1. Performance and Safety of Existing Structures: Introduction, Høj, Niels Peter; Söderqvist, Marja-Kaarina, SEI, Vol. 19, No. 2, May 2009, pp. 165.
2. Reuse Dismantling and Performance Evaluation of Reusable Members, Fujita, Masanori; Iwata, Mamoru, SEI, Vol. 18, No. 3, Aug. 2008, pp. 230–237.
3. Performance-Based Seismic Retrofit Design of a Long-Span Truss Bridge—Minato Bridge—Using New Control Technologies, Kanaji, Hidesada; Fujino, Yozo; Watanabe, Eiichi, SEI, Vol. 18, No. 3, Aug. 2008, pp. 271–277.
4. Field Performance of a Composite Log-Concrete Bridge, de Góes, Jorge Luís Nunes; Júnior, Calil, SEI, Vol. 18, No. 1, Feb. 2008, pp. 79–83.
5. Performance-based Standard Specifications for Maintenance and Repair of Concrete Structures in Japan, Ueda, Tamon; Takewaka, Koji, SEI, Vol. 17, No. 4, Nov. 2007, pp. 359–366.
6. Time-Variant Structural Performance of the Certosa Cable-Stayed Bridge, Biondini, Fabio; Frangopol, Dan M.; Malerba, Pier Giorgio, SEI, Vol. 16, No. 3, Aug. 2006, pp. 235–244.
7. Life-Cycle Performance of Deteriorating Structures, Adey, Bryan; Brühwiler, Eugene; Frangopol, Dan M.; Faber, Michael H., SEI, Vol. 13, No. 3, Aug. 2003, pp. 202–204.
8. Increasing Knowledge of Structural Performance, Smith, Ian F. C., SEI, Vol. 11, No. 3, Aug. 2001, pp. 191–195.

(12) Robustness

1. Load Capacity Evaluation of Existing Railway Bridges based on Robustness Quantification, Wisniewski, Dawid; Casas, Joan R.; Ghosn, Michel, SEI, Vol. 16, No. 2, May 2006, pp. 161–166.

(13) Safety

1. Assessment of the Seismic Resistance and Structural Safety of Existing Multistory Residential Buildings, Schwarz, Stephan; Leibovich, Edward; Yankelevsky, David Z., SEI, Vol. 19, No. 2, May 2009, pp. 165–173.
2. Crash Safety of Lightweight Gantry in Aluminium, Heglund, Kim; Czujko, Jurek, SEI, Vol. 16, No. 4, Nov. 2006, pp. 363–366.

3. Safety of Laminated Glass Structures after Initial Failure, Kott, Alexander; Vogel, Thomas, SEI, Vol. 14, No. 2, May 2004, pp. 134–138.
4. Safety - A Matter of Risk, Cost and Consensus, Schneider, Jörg, SEI, Vol. 10, No. 4, Nov. 2000, pp. 266–269.

Group (D):

(14) Corrosion

1. Durability of Titanium-Clad Steel Plates used as an Anti-Corrosion System, Nakamura, Shun-ichi; Homma, Koji, SEI, Vol. 10, No. 4, Nov. 2000, pp. 262–265.
2. Corrosion Mechanism and Protection Methods for Suspension Bridge Cables, Furuya, Kazuhiko; Kitagawa, Makoto; Nakamura, Shun-ichi; Suzumura, Keita, SEI , Vol. 10, No. 3, Aug. 2000, pp. 189–193.
3. Corrosion-resistant Fastenings in Road Tunnels—Field Tests, Böhni, Hans; Haselmair, Helmut; Übeleis, Andreas M., SEI , Vol. 2, No. 4, Nov. 1992, pp. 253–258.
4. Corrosion Measurement on Bridge Decks, Potential Mapping, Marquardt, Helmut, SEI , Vol. 1, No. 3, Aug. 1991, pp. 49–50.

(15) Cracking

1. Crack Width Evaluation of Reinforced Concrete Membrane Elements, Giordano, Luca; Mancini, Giuseppe, SEI , Vol. 19, No. 3, Aug. 2009, pp. 256–261.
2. Concrete Cracking in Composite Bridges: Tests, Models and Design Proposals, Navarro, Miguel Gómez; Lebet, Jean-Paul, SEI, Vol. 11, No. 3, Aug. 2001, pp. 184–190.
3. Tension Chord Model for Structural Concrete, Marti, Peter; Alvarez, Manuel; Kaufmann, Walter; Sigrist, Viktor, SEI, Vol. 8, No. 4, Nov. 1998, pp. 287–298.
4. Retrofit of Fatigue Cracks in Elevated Structures, Irshad, Mohammad; Reed, Timothy, SEI , Vol. 3, No. 3, Aug. 1993, pp. 178–180.

(16) Damage

1. A New Structural System Using Innovative High-Strength Steel Aiming at Zero Earthquake Damage, Takanashi, Koichi; Miyazaki, Ken-ichi; Yamazaki, Kenji; Shimura, Yasumi, SEI, Vol. 20, No. 1, Feb. 2010, pp. 66–71.
2. Reconnaissance and Preliminary Observations of Bridge Damage in the Great Wenchuan Earthquake, China, Liu, Zhao, SEI, Vol. 19, No. 3, Aug. 2009, pp. 277–282.
3. Wind-Induced Damages to a Three-Span, Continuous, Concrete Arch Bridge under Construction, Ge, Yao-Jun; Yang, Yong-Xin; Pang, Jia-Bin; Xiang, Hai-Fan, SEI, Vol. 17, No. 2, May 2007, pp. 141–150.
4. Assessment of Damaged Post-Tensioning Tendons, Vill, Markus; Eichinger, Eva M.; Kollegger, Johann, SEI, Vol. 16, No. 1, Feb. 2006, pp. 44–48.
5. Damage Analysis for a Fire Exposed Industrial Building, Peker, Kerem; Pekmezci, Bekir, SEI, Vol. 13, No. 4, Nov. 2003, pp. 245–248.
6. Structural Damage and Remedial Measures for the Temples of Angkor, Cambodia, Croci, Giorgio, SEI, Vol. 11, No. 4, Nov. 2001, pp. 234–236.
7. Glue Joints Used For Reinforcing the Damaged Ends of Wooden Beams, Jasienko, Jerzy, SEI, Vol. 11, No. 4, Nov. 2001, pp. 246–250.
8. Damage Assessment and Analysis of the 1997 Umbria-Marche Earthquakes, Spence, Robin; D'Ayala, Dina, SEI, Vol. 9, No. 3, Aug. 1999, pp. 229–233.
9. Reconstructing an Earthquake-Damaged Building, Kobe, Uchida, Naoki; Takagaki, Toshio, SEI, Vol. 8, No. 3, Aug. 1998, pp. 179–180.
10. Damages to Steel Structures Caused, the 1995 Kobe Earthquake, Horikawa, Kousuke; Sakino, Yoshihiro, SEI, Vol. 6, No. 3, Aug. 1996, pp. 181–182.

11. Damage Incurred at the Bridges over the Inn River at Kufstein (Austria), and their Repair, Wicke, Manfred, SEI, Vol. 1, No. 1, Feb. 1991, pp. 28–34.

(17) Deterioration

1. Deterioration of Reinforced Concrete Structures in a Marine Environment, Mallick, D. V.; Tawil, M. M., SEI, Vol. 2, No. 3, Aug. 1992, pp. 190–193.
2. Life-Cycle Performance of Deteriorating Structures, Adey, Bryan; Brühwiler, Eugene; Frangopol, Dan M.; Faber, Michael H., SEI, Vol. 13, No. 3, Aug. 2003, pp. 202–204.

(18) Fatigue

1. Development and Application of a Fatigue Class Catalogue for Riveted Bridge Components, Taras, Andreas; Greiner, Richard, SEI, Vol. 20, No. 1, Feb. 2010, pp. 91–103.
2. Fatigue Safety Assessment of Existing Steel Bridges in China, Wang, Chunsheng; Chen, Weizhen; Chen, Airong, SEI, Vol. 19, No. 2, May 2009, pp. 174–179.
3. Evaluation of Fatigue Category of Riveted Steel Bridge Connections, Matar, Ehab B., SEI, Vol. 17, No. 1, Feb. 2007, pp. 72–78.
4. Fatigue of Aluminium Bridge Decks, Maljaars, Johan; Soetens, Frans; van Straalen, Ijsbrand, SEI, Vol. 16, No. 4, Nov. 2006, pp. 305–311.
5. Fatigue Tests for a Riveted Steel Railway Bridge in Salzburg, Matar, Ehab Boghdadi; Greiner, Richard, SEI, Vol. 16, No. 3, Aug. 2006, pp. 252–260.
6. Fatigue Strength of Glass Panels under Cyclic Loading, Boxheimer, Kaja; Wörner, Johann-Dietrich, SEI, Vol. 14, No. 2, May 2004, pp. 126–128.
7. Analysis of Multi-Beam Steel Bridges for Fatigue, Corwin, Eric; Dexter, Robert, SEI, Vol. 12, No. 4, Nov. 2002, pp. 249–254.
8. Fatigue Cracking in a Steel Railway Bridge, Åkesson, Björn; Edlund, Bo; Shen, Dafeng, SEI , Vol. 7, No. 2, May 1997, pp. 118–120.
9. Fatigue of Thin-Walled Plate Girders, Okura, Ichiro; Yen, Ben T.; Fisher, John W., SEI , Vol. 3, No. 1, Feb. 1993, pp. 39–44.
10. Length Effect on Fatigue of Wires and Strands: IABSE Workshop, El Paular (Madrid), Spain, September 23–25, 1992, Canteli, Alfredo Fernandez, SEI , Vol. 3, No. 1, Feb. 1993, pp. 57.
11. Retrofit of Fatigue Cracks in Elevated Structures, Irshad, Mohammad; Reed, Timothy, SEI , Vol. 3, No. 3, Aug. 1993, pp. 178–180.
12. Fatigue Tests of Monorail Guideway Girders, Okura, Ichiro, SEI , Vol. 1, No. 2, May 1991, pp. 28–36.

(19) Fracture

1. Fracture Critical Study of an Historic Wrought Iron Bridge, Badoux, Marc; Sparks, Patrick, SEI, Vol. 8, No. 2, May 1998, pp. 136–139.
2. Brittle Fracture Failure Due to Deficient Welding, Kaminetzky, Dov, SEI, Vol. 3, No. 4, Nov. 1993, pp. 258–259.

(20) Scour

1. Bridge Scour Failuress**,** Mallick, D. V.; Tawil, M. M., SEI, Vol. 4, No. 1, Feb. 1994, pp. 44–46.

(21) Vibration

1. Wind Vibration Reduction at Alicante Airport Control Tower, Spain, Astiz, Miguel A.; Ortega, Luis M., Vol. 15, No. 4, Nov. 2005, pp. 236–239.
2. Pedestrian Lateral Action on lively Footbridges: A New Load Model, Venuti, Fiammetta; Bruno, Luca; Napoli, Paolo, SEI, Vol. 17, No. 3, Aug. 2007, pp. 236–241.
3. Wind-Induced Vibration of Tubular Diagonals of the Werkspoorbridge, Vrouwenvelder, A. C. W. M.; Hoeckman, W., Vol. 14, No. 4, Nov. 2004, pp. 314–321.

4. Lateral Vibration on a Pedestrian Cable-Stayed Bridge, Nakamura, Shun-Ichi; Fujino, Yozo, Vol. 12, No. 4, Nov. 2002, pp. 295–300.
5. Vibration-Isolation Retrofit for the National Museum of Western Art, Tokyo, Ito, Akihiro; Nakagawa, Ryugo; Abe, Fumiaki; Nakamura, Yasukazu, Vol. 8, No. 3, Aug. 1998, pp. 181–182.
6. Advances in Ambient Vibration Testing: Ganter Bridge, Switzerland, Felber, Andreas; Cantieni, Reto, Vol. 6, No. 3, Aug. 1996, pp. 187–190.
7. Tuned Vibration Absorbers for "Lively" Structures, Bachmann, Hugo; Weber, Benedikt, Vol. 5, No. 1, Feb. 1995, pp. 31–36.
8. Suppression of Wind-Induced Vibration of Tower-Shaped Structures, Dynamic Dampers, Ueda, Toshio; Nakagaki, Ryoji; Koshida, Kazuo, Vol.3, No. 1, Feb. 1993, pp. 50–53.
9. Vibration Upgrading of Gymnasia, Dance Halls and Footbridges, Bachmann, Hugo, Vol.2, No. 2, May 1992, pp. 118–124.

(22) Weathering

1. Weathering Steel in Bridges, Fischer, Manfred, SEI, Vol. 5, No. 1, Feb. 1995, pp. 51–54.

Group (E):

(23) Prevention

1. Reasons and prevention of corrosion-induced failures of prestressing steel in concrete, Nuernberger, U., SEI, Vol. 1, No. 1, Dec. 2009, pp. 29–39
2. Preventive Medicine: Condition Assessment, Ratay, Robert T., SEI, Vol. 17, No. 2, May 2007, pp. 113–113.

(24) Protection

1. Rockfall Protection: Introduction, Vogel, Thomas; Høj, Niels Peter, SEI, Vol. 19, No. 3, Aug. 2009, pp. 304.
2. An Uncoupled Approach for the Design of Rockfall Protection Tunnels, Calvetti, Francesco; di Prisco, Claudio, SEI, Vol. 19, No. 3, Aug. 2009, pp. 342–347.
3. Bridge Protection of the Rosario-Victoria Bridge, Argentina, Saul, Reiner; Humpf, Karl; Patsch, Armin, SEI, Vol. 13, No. 4, Nov. 2003, pp. 227–231.
4. Protective Structure for Ancient Ruins at Ephesus, Turkey, Ziesel, Wolfdietrich, SEI, Vol. 10, No. 4, Nov. 2000, pp. 235–237.
5. Corrosion Mechanism and Protection Methods for Suspension Bridge Cables, Furuya, Kazuhiko; Kitagawa, Makoto; Nakamura, Shun-ichi; Suzumura, Keita, SEI, Vol. 10, No. 3, Aug. 2000, pp. 189–193.

Group (F):

(25) Preservation

1. Preservation of Historic Bridges in New South Wales, Australia, Maxwell, Ken, SEI, Vol. 13, No. 2, May 2003, pp. 133–136.
2. Main Cable Preservation for the Williamsburg Bridge, Bruschi, Maria Grazia; Koglin, Terry L., SEI, Vol. 4, No. 2, May 1994, pp. 122–124.
3. Structural Preservation of Italy's Monuments, Finzi, Leo, SEI, Vol. 3, No. 1, Feb. 1993, pp. 6.
4. Structural Preservation of the Fraccaro Tower, Pavia, Ballio, Giulio, SEI, Vol. 3, No. 1, Feb. 1993, pp. 9–11.
5. Structural Preservation of the Historical Center of Naples, Mandara, Alberto; Mazzolani, Federico M., SEI, Vol. 3, No. 1, Feb. 1993, pp. 11–14.
6. Preservation of the Duomo of Milan, da Passano, Carlo Ferrari, SEI, Vol. 3, No. 1, Feb. 1993, pp. 17–19.

7. Preservation of San Fruttuoso Abbey, Camogli, Castoldi, Aldo; Rossi, Pier Paolo, SEI, Vol. 3, No. 1, Feb. 1993, pp. 28–29.

(26) Refurbishment

1. Attainment of an Integral Abutment Bridge through the Refurbishment of a Simply Supported Structure, Zordan, Tobia; Briseghella, Bruno, SEI, Vol. 17, No. 3, Aug. 2007, pp. 228–234.
2. Aluminium Structures in Refurbishment: Case of the Real Ferdinando Bridge on Garigliano River, Mazzolani, Federico M., SEI, Vol. 16, No. 4, Nov. 2006, pp. 352–355.

(27) Rehabilitation

1. Rehabilitation of the Hsi-Lo Bridge in Taiwan, Pu, Jun-Ping; Wu, Jui-Loung; Lin, Chia-Hung; Lin, Chia-Chi, SEI, Vol. 17, No. 3, Aug. 2007, pp. 246–251.
2. Rehabilitation of the Figueira da Foz Bridge, Rito, Armando; Appleton, Julio, SEI, Vol. 15, No. 2, May 2005, pp. 92–92.
3. Failure of Hollow Ribbed Slabs in Buildings - Suggested Rehabilitation Method, Pisanty, Avraham; Mairanz, Benjamin, SEI, Vol. 14, No. 1, Feb. 2004, pp. 46–49.
4. Assessment and Rehabilitation of Steel Bridges: An Introduction, Ghoshal, Amitabha, SEI, Vol. 12, No. 4, Nov. 2002, pp. 249.
5. Experiences in Rehabilitation of Steel Bridges, Ghosh, Utpal; Ghoshal, Amitabha, SEI, Vol. 12, No. 4, Nov. 2002, pp. 269–272.
6. Rehabilitation of Tornionjoki Steel Truss Bridge, Finland, Kääriäinen, Juha; Pulkkinen, Pekka, SEI, Vol. 12, No. 4, Nov. 2002, pp. 273–275.
7. Rehabilitation of the Banatul Museum, Timisoara, Romania, Bob, Corneliu; Dan, Sorin; Balut, Nicolae; Gadea, Angelu, SEI, Vol. 11, No. 4,Nov. 2001, pp. 231–233.
8. Rehabilitation of St. Ursanne Railway Viaduct, Switzerland, Voutaz, Bertrand; Reber, Jean-Jacques, SEI, Vol. 11, No. 4, Nov. 2001, pp. 241–243
9. Rehabilitation of the Matsesta River Bridge, Russian Fed., Blinkov, Leonid S.; Valiev, Sherali N., SEI, Vol. 11, No. 3, Aug. 2001, pp. 181–183.
10. Rehabilitation of Sidi M'Cid Suspension Bridge, Algeria, Petrangeli, Marco; Petrangeli, Marcello, SEI, Vol. 10, No. 4, Nov. 2000, pp. 254–258.
11. Rehabilitation of the Highest Wooden Tower in Poland, Ajdukiewicz, Andrzej; Brol, Janusz; Malczyk, Andrzej; Wlaszczuk, Marek, SEI, Vol. 10, No. 3, Aug. 2000, pp. 161–163.
12. Rehabilitation of the Salginatobel Bridge, Figi, Heinrich, SEI, Vol. 10, No. 1, Feb. 2000, pp. 21–23.
13. Rehabilitation of the Devil's Bridge over Schollenen Gorge, Huber, Heribert, SEI, Vol. 10, No. 1, Feb. 2000, pp. 23–25.
14. Rehabilitation of Two Overpasses on a Russian Highway, Blinkov, L.; Veinblat, B.; Johnson, B.; Reshetnikov, V., SEI, Vol. 10, No. 1, Feb. 2000, pp. 54–55.
15. Softening Instead of Strengthening for Seismic Rehabilitation, Bachmann, Hugo; Wenk, Thomas, SEI, Vol. 10, No. 1, Feb. 2000, pp. 61–65.
16. Rehabilitation of a Highway Bridge in Service, Austria, Unterholzner, Peter; Kirsch, Peter; Obholzer, Anton; Brunnsteiner, Wilhelm; Geisler, Peter, SEI, Vol. 7, No. 2, May 1997, pp. 82–84.
17. Rehabilitation of the Mexico City Cathedral, Meli, Roberto; Sanchez-Ramirez, A. Roberto, SEI, Vol. 7, No. 2, May 1997, pp. 101–106.
18. Rehabilitation of a Grain Silo Complex, Iran, Ghalibafian, Mehdi; Azarin, Ramin, SEI, Vol. 7, No. 3, Aug. 1997, pp. 184.
19. Rehabilitation of Two Floating Bridges, Turkey, Arda, Tevfik Seno; Yardimci, Nesrin; Eyrekci, Oguz, SEI, Vol. 6, No. 1, Feb. 1996, pp. 17–18.

(28) Remedial

1. Structural Damage and Remedial Measures for the Temples of Angkor, Cambodia, Croci, Giorgio, SEI, Vol. 11, No. 4, Nov. 2001, pp. 234–236.

2. Timber Decay: Causes and Remedies, Singh, Jagjit; Mann, Peter; White, Nia, SEI, Vol. 6, No. 3, Aug. 1996, pp. 200–202.

(29) Renew

1. Renewing the Reichstag, Andrä, Hans-Peter; Fink, Roland, SEI, Vol. 7, No. 4, Nov. 1997, pp. 245–248.

(30) Repair

1. Assessment, Repair and Widening of the Villeneuve-Loubet Bridge, France, Vion, Philippe; Poineau, Daniel, SEI, Vol. 18, No. 4, Nov. 2008, pp. 343–345.
2. An Introduction: Operations, Maintenance and Repair of Structures, Sobrino, Juan A., SEI, Vol. 17, No. 4, Nov. 2007, pp. 328.
3. Performance-based Standard Specifications for Maintenance and Repair of Concrete Structures in Japan, Ueda, Tamon; Takewaka, Koji, SEI, Vol. 17, No. 4, Nov. 2007, pp. 359–366.
4. Optimum Lifetime Planning of Bridge Inspection and Repair Programs, Frangopol, Dan M.; Estes, Allen C., SEI, Vol. 9, No. 3, Aug. 1999, pp. 219–223.
5. Repair and Strengthening of New York's Infrastructure, Englot, Joseph M., SEI, Vol. 6, No. 2, May 1996, pp. 102–106.
6. Testing and Repair of Concrete Silos, Ajdukiewicz, Andrzej, SEI, Vol. 6, No. 4, Nov. 1996, pp. 278–281.
7. Repair and Strengthening of Structures, Pakvor, Aleksandar, SEI, Vol. 5, No. 2, May 1995, pp. 70.
8. Repair and Strengthening of Concrete Structures: General Aspects, Pakvor, Aleksandar, SEI, Vol. 5, No. 2, May 1995, pp. 70–73.
9. New Repair Techniques for Concrete Structures, Iványi, György; Buschmeyer, Wilhelm, SEI, Vol. 5, No. 2, May 1995, pp. 85–87.
10. What to Learn from the Öland Bridge Repair?, Ingvarsson, Hans, SEI, Vol. 5, No. 2, May 1995, pp. 88–89.
11. Decision Criteria for Concrete Bridge Repair, Branco, Fernando A.; de Brito, José, SEI, Vol.5, No. 2, May 1995, pp. 92–95.
12. Collapse of a Warehouse and Its Repair, Kratena, Jindrich; Krejci, Hanus, SEI, Vol. 5, No. 3, Aug. 1995, pp. 192–193.
13. Seismic Repair and Upgrading of a Dome Lantern in Assisi, Menegotto, Marco, SEI, Vol.3, No. 1, Feb. 1993, pp. 34–36.
14. Experience with Hydrodemolition for Bridge Repair, Eriksson, Bosse, SEI, Vol. 2, No. 3, Aug. 1992, pp. 205–206.
15. Damage Incurred at the Bridges over the Inn River at Kufstein (Austria), and their Repair, Wicke, Manfred, SEI, Vol. 1, No. 1, Feb. 1991, pp. 28–34.

(31) Replacement of Components

1. Replacement of Bearings in the Golden Horn Bridge, Turkey, Yanagihara, Masahiro; Matsuzawa, Toshimitsu; Kudo, Mizsuhiro; Turker, Turgay, SEI, Vol. 10, No. 2, May 2000, pp. 121–123.

(32) Restoration

1. Restoration of Historical Structures: An Introduction, Croci, G., SEI, Vol. 11, No. 4, Nov. 2001, pp. 222.
2. Restoration of the San Jacinto Monument, USA, Koerber, Jeffrey; Hunderman, Harry J.; Paulson, Conrad, SEI, Vol. 11, No. 4, Nov. 2001, pp. 227–230.
3. Restoration of the Railway Station Roofs of 's-Hertogenbosch, Netherlands, Vákár, László I., SEI, Vol. 11, No. 4, Nov. 2001, pp. 236–240.
4. Structural Restoration of Historic Buildings: General Guidelines, Croci, Giorgio, SEI, Vol. 5, No. 2, May 1995, pp. 74–75.

5. Restoration of the Zisa Palace, Palermo, Zingone, Gaetano, SEI, Vol. 3, No. 1, Feb. 1993, pp. 22–25.
6. Restoration of a 100 Year Old Iron Bridge, Paderno, Nascé, Vittorio, SEI, Vol. 3, No. 1, Feb. 1993, pp. 37–38.

(33) Retrofit

1. Retrofitting a Short Span Bridge with a Semi-Integral Abutment Bridge: The Treviso Bridge, Russo, Gaetano; Bergamo, Otello; Damiani, Luigi, SEI, Vol. 19, No. 2, May 2009, pp. 137–141.
2. Mountain Pass Slope Failure Retrofitted with a Half Viaduct Bridge Structure, South Africa, Kruger, E.J.; Newmark, A.A.; Smuts, M., SEI, Vol. 18, No. 4, Nov. 2008, pp. 318–322.
3. Performance-Based Seismic Retrofit Design of a Long-Span Truss Bridge Minato Bridge-Using New Control Technologies, Kanaji, Hidesada; Fujino, Yozo; Watanabe, Eiichi, SEI, Vol. 18, No. 3, Aug. 2008, pp. 271–277.
4. Retrofitting of Tautendorf Valley Bridge in Highway A9 Berlin, Germany, Reintjes, K.H.; Wolf, H., SEI, Vol. 15, No. 3, Aug. 2005, pp. 151.
5. Extending Service Life through Pre-emptive Retrofits, Fisher, John W.; Connor, Robert J., SEI, Vol. 12, No. 4, Nov. 2002, pp. 264–268.
6. Seismic Evaluation and Retrofit of Historical Churches, Sorace, Stefano; Terenzi, Gloria, SEI, Vol. 12, No. 4, Nov. 2002, pp. 283–288.
7. Strengthening, Retrofitting and Upgrading of Timber Structures with High-Strength Fibres, Kropf, François W.; Meierhofer, Ulrich, SEI, Vol. 10, No. 3, Aug. 2000, pp. 178–181.
8. Vibration-Isolation Retrofit for the National Museum of Western Art, Tokyo, Ito, Akihiro; Nakagawa, Ryugo; Abe, Fumiaki; Nakamura, Yasukazu, SEI, Vol. 8, No. 3, Aug. 1998, pp. 181–182.
9. Seismic Retrofit of the Oakland City Hall, Oakland, California, Elsesser, Eric; Honeck, William; Walters, Mason, SEI, Vol. 5, No. 1, Feb. 1995, pp. 12–14.
10. Segmental Aerial Structure Retrofit and Load Testing, Irshad, Mohammad; Poston, Randall W., SEI, Vol. 3, No. 2, May 1993, pp. 108–112.
11. Retrofit of Fatigue Cracks in Elevated Structures, Irshad, Mohammad; Reed, Timothy, SEI, Vol. 3, No. 3, Aug. 1993, pp. 178–180.

(34) Stiffening

1. Strengthening, Stiffening and Repair of Concrete Structures, R.F.Warner, IABSE PERIODICA 2/1981 No. S-17/81, p. 25–43.

(35) Strengthening

1. Strengthening Bridges, Developing Composite Action in Existing Non-Composite Bridge Girders, Kwon, Gunup; Engelhardt, Michael D.; Klingner, Richard E., SEI, Vol. 19, No. 4, Nov. 2009, pp. 432–437.
2. Strengthening the Röslautal Bridge Using Innovative Techniques, Germany, Blaschko, Michael; Zehetmaier, Gerhard, SEI, Vol. 18, No. 4, Nov. 2008, pp. 346–350.
3. Strengthening Coalport Bridge, De Voy, Julian; Williams, John M., SEI, Vol. 17, No. 2, May 2007, pp. 178–183.
4. Fiber-Reinforced Composites for the Strengthening of Masonry Structures, Nanni, Antonio; Tumialan, J. Gustavo, SEI, Vol. 13, No. 4, Nov. 2003, pp. 271–278.
5. Fiber-Reinforced Polymer for Structural Strengthening: Post-Tensioning of Steel Silos, De Lorenzis, Laura; Micelli, Francesco; Tegola, Antonio La, SEI, Vol. 13, No. 2, May 2003, pp. 124–127.
6. Strengthening of an Old Arch Truss Bridge, Austria, Holzinger, Helmut; Jeschko, Andreas; Robra, Jörgen; Ramberger, Günter, SEI, Vol. 12, No. 4, Nov. 2002, pp. 276–280.
7. Effectiveness of FRP for Strengthening Concrete Bridges, Rizkalla, Sami; Hassan, Tarek, SEI, Vol. 12, No. 2, May 2002, pp. 89–95.

8. Strengthening of a Concrete Bridge and Loading to Failure, Vogel, Thomas; Ulaga, Tomaz, SEI, Vol. 12, No. 2, May 2002, pp. 105–110.
9. Strengthening of the West Gate Bridge Approach Span, Melbourne, Gosbell, Tim, SEI, Vol. 12, No. 1, Feb. 2002, pp. 14–16.
10. Strengthening the Basilica of St Francis of Assisi after the September 1997 Earthquake, Croci, Giorgio, SEI, Vol. 11, No. 3, Aug. 2001, pp. 207–210.
11. Strengthening, Retrofitting and Upgrading of Timber Structures with High-Strength Fibres, Kropf, François W.; Meierhofer, Ulrich, SEI, Vol. 10, No. 3, Aug. 2000, pp. 178–181.
12. Softening Instead of Strengthening for Seismic Rehabilitation, Bachmann, Hugo; Wenk, Thomas, SEI, Vol. 10, No. 1, Feb. 2000, pp. 61–65.
13. Strengthening a High-Rise for a Rooftop Helipad, Bangkok, Buddee, Samard, SEI, Vol. 8, No. 2, May 1998, pp. 88.
14. Strengthening a Bridge with Advanced Materials, Walser, Rolf; Steiner, Werner, SEI, Vol. 7, No. 2, May 1997, pp. 110–112.
15. Repair and Strengthening of New York's Infrastructure, Englot, Joseph M., SEI, Vol. 6, No. 2, May 1996, pp. 102–106.
16. Repair and Strengthening of Structures, Pakvor, Aleksandar, SEI, Vol. 5, No. 2, May 1995, pp. 70.
17. Repair and Strengthening of Concrete Structures: General Aspects, Pakvor, Aleksandar, SEI, Vol. 5, No. 2, May 1995, pp. 70–73.
18. Strengthening Bridges with Unbonded Prestressing, Falkner, Horst; Huang, Zhen; Teutsch, Manfred, SEI, Vol. 5, No. 2, May 1995, pp. 76–77.
19. Bridge Strengthening with Additional Prestressing, Straninger, Walter; Wicke, Manfred, SEI, Vol. 5, No. 2, May 1995, pp. 78–80.

(36) Upgrading

1. Priorities in Earthquake Upgrading of Existing Structures, Kölz, Ehrfried; Bürge, Marcel, SEI, Vol. 11, No. 3, Aug. 2001, pp. 202–206.
2. Strengthening, Retrofitting and Upgrading of Timber Structures with High-Strength Fibres, Kropf, François W.; Meierhofer, Ulrich, SEI, Vol. 10, No. 3, Aug. 2000, pp. 178–181.
3. Upgrading the South Terrace, Arsenal Football Grounds, London, Bardhan-Roy, B. K., SEI, Vol. 4, No. 1, Feb. 1994, pp. 11–13.
4. Seismic Repair and Upgrading of a Dome Lantern in Assisi, Menegotto, Marco, SEI, Vol. 3, No. 1, Feb. 1993, pp. 34–36.
5. Vibration Upgrading of Gymnasia, Dance Halls and Footbridges, Bachmann, Hugo, SEI, Vol. 2, No. 2, May 1992, pp. 118–124.

(37) Widening

1. Widening of the Penang Bridge, Malaysia, Buckby:, Roger; Peng, Chen Wai; Corbett, Paul; Singh, Muhinder, SEI, Vol. 19, No. 1, Feb. 2009, pp. 41–45.
2. Analysis of Structural Behaviour in Widened Concrete Box Girder Bridges, Shi, Xuefei; Li, Xiaoxiang; Ruan, Xin; Ying, Tianyi, SEI, Vol. 18, No. 4, Nov. 2008, pp. 351–355.
3. Widening of Bridges: Introduction, Buckby:, Roger, SEI, Vol. 18, No. 4, Nov. 2008, pp. 314.
4. Widening of the Cable-Stayed Bridge over the Rande Strait, Spain, Calzón, Julio Martínez; Vilardell, Manuel Juliá; Corral, Álvaro Serrano; Navarro, Miguel Gómez, SEI, Vol. 18, No. 4, Nov. 2008, pp. 314–317.
5. Design and Experimental Investigation of the Joints of Inclined Struts for the Widening of Bridge Deck Slabs, Menétrey, Philippe; Brühwiler, Eugen, SEI, Vol. 18, No. 4, Nov. 2008, pp. 337–342.
6. Assessment, Repair and Widening of the Villeneuve-Loubet Bridge, France, Vion, Philippe; Poineau, Daniel, SEI, Vol. 18, No. 4, Nov. 2008, pp. 343–345.
7. Design Solutions for Widening the A1–A9–A14 Italian Highways, Furlanetto, Guido; Torricelli, Lucio Ferretti; Marchiondelli, Alessandra, SEI, Vol. 18, No. 4, Nov. 2008, pp. 356–364.

8. An International Perspective: Widening Existing Bridges with Orthotropic Steel Deck Panels, Huang, Carl; Mangus, Alfred R., SEI, Vol. 18, No. 4, Nov. 2008, pp. 381–389.
9. Laterally Cantilevered Space Frame for the Roadway Widening in Steep-Sloped Mountainous Areas, Zhou, Zhixiang; Li, Fang; Qujue, Awang, SEI, Vol. 18, No. 3, Aug. 2008, pp. 254–258.
10. Reconstruction of Masonry Structures during the Widening of the Federal Motorways in the New German Countries, Reintjes, Karl - Heinz, SEI, Feb. 2007, pp. 44–49.
11. Widening of the Elche de la Sierra Arch Bridge, Spain, Tanner, Peter; Bellod, Juan Luis, SEI, Vol. 15, No. 3, Aug. 2005, pp. 148.
12. Widening of the Aare Bridges at Ruppoldingen, Dauner, Hans G.; Schibli, Hansjörg; Stucki, Dieter, SEI, Vol. 10, No. 1, Feb. 2000, pp. 26–27.

Group (G):

(38) Rebuilding

1. Rebuilding the Taggenbrunn Railway Bridge, Austria, Hillbrand, Herfried, SEI, Vol. 7, No. 1, Feb. 1997, pp. 14–16.

(39) Reconstruction

1. Reconstruction of Masonry Structures during the Widening of the Federal Motorways in the New German Countries, Reintjes, Karl - Heinz, SEI, Vol. 17, Feb. 2007, pp. 44–49.
2. Reconstruction of an Office Building, Roosevelt Square, Budapest, Almasi, Jozsef; Nemes, Balint, SEI, Vol. 16, No. 1, Feb. 2006, pp. 24–27.
3. Prestressing Work in Reconstruction of the Dresden Frauenkirche Cupola, Germany, Dornacher, Siegfried; Schaeffer, Ernst, SEI, Vol. 13, No. 4, Nov. 2003, pp. 264–267.
4. Reconstruction of a Bridge in Vyborg, Russian Fed., Blinkov, Leonid; Zhurbin, Alexej; Surovtsev, Alexej, SEI, Vol. 10, No. 2, May 2000, pp. 102–103.
5. Reconstruction of an Ice Hockey Hall in Kosice, Slovakia, Kanocz, Jan D.; Kmet', Stanislav, SEI, Vol. 9, No. 2, May 1999, pp. 116–118.
6. Reconstruction of the Moscow Ring Road, Russia, Perevoznikov, Boris F.; Seliverstov, Vadim A., SEI, Vol. 9, No. 2, May 1999, pp. 137–142.
7. Reconstructing Ikuta Shrine with Composite Columns, Nagano, Yasuyuki; Okamoto, Tatsuo, SEI, Vol. 8, No. 3, Aug. 1998, pp. 175–176.
8. Reconstructing the Oldest Western-Style Building in Kobe, Imai, Shigeyuki; Suzuki, Naomiki, SEI, Vol. 8, No. 3, Aug. 1998, pp. 177–178.
9. Reconstructing an Earthquake-Damaged Building, Kobe, Uchida, Naoki; Takagaki, Toshio, SEI, Vol. 8, No. 3, Aug. 1998, pp. 179–180.

(40) Replacement of Structures

1. US Grant Bridge Replacement, Kumarasena, Sena; McCabe, Ray, SEI, Vol. 18, No. 1, Feb. 2008, pp. 56–61.
2. Acosta Bridge Replacement, Jacksonville, Florida, Pielstick, Brett H., SEI, Vol. 5, No. 1, Feb. 1995, pp. 19–20.
3. Replacing Historic Rail Bridge in the Beer Sheba Valley, Israel, Shamir, Eliezer, SEI, Vol. 15, No. 4, Nov. 2005, pp. 216–218.

Group (H):

(41) Reuse

1. Reuse Dismantling and Performance Evaluation of Reusable Members, Fujita, Masanori; Iwata, Mamoru, SEI, Vol. 18, No. 3, Aug. 2008, pp. 230–237.

2. Design for Dismantling and Reuse of an Exhibition Pavilion, Germany, Tanner, Peter; Thomas, Juan Luis Bellod, SEI, Vol. 11, No. 2, May 2001, pp. 116–119.

(II) IABSE Conferences

Many papers can be found in IABSE Conference proceedings on topics related to SED 12: (rehabilitation, repair, retrofit, strengthening, upgrading, conservation, . . . of structures). The following are listed as examples, or as specialized conferences:
http://www.iabse.org/publications/iabsereports/index.php

1. Codes in Structural Engineering, Developments and Needs for International Practice, IABSE *–fib* Conference, Dubrovnik, Croatia, May 3–5, 2010. Part 5: Existing Structures and Maintenance, Session (5.1) Codes and General management Procedures; (5.2) Updating Information & Adapted Load Models; (5.3) Rehabilitation, Strengthening & Upgrading; (5.4) Comparisons with Codes.
2. Sustainable Infrastructure—Environment Friendly, Safe and Resource Efficient, IABSE Symposium, Bangkok, Thailand, September 9–11, 2009. Session: Operation, Monitoring, Maintenance and Repair.
3. Creating and Renewing Urban Structures, 17th Congress of IABSE Chicago, USA, September 17–19, 2008. Session (1B): Strengthening and Upgrading of Buildings; Session (3C): Strengthening and Upgrading of Bridges; Session (6C): Rehabilitation and Replacement of Bridges.
4. Improving Infrastructure Worldwide, IABSE Symposium, Weimar, Germany, September 19–21, 2007.
5. Responding to Tomorrow's Challenges in Structural Engineering, IABSE Symposium, Budapest, Hungary, September 13–15, 2006. Poster Session: Refurbishment, Repair and strengthening; Session: Refurbishment, Repair and Strengthening.
6. Operation, Maintenance and Rehabilitation of Large Infrastructure Projects, Bridges and Tunnels, IABSE Conference, Copenhagen, Denmark, May 15–17, 2006.
7. Structures and Extreme Events, IABSE Symposium, Lisbon, Portugal, September 14–17, 2005. Session: Structural Evaluation and Monitoring.
8. Metropolitan Habitats and Infrastructure, IABSE Conference, Shanghai, China, September 22–24, 2004. Session (4B): Maintenance, Operation and Life Cycle Considerations. Session (4S) Maintenance, Operation and Life Cycle Consideration of Structures.
9. Towards a Better Built Environment—Innovation, Sustainability, Information Technology, IABSE Symposium, Melbourne, 8–13 September 2002. Session: Strengthening and Repair, & Session: FRP Strengthening and Repair.
10. Saving Buildings in Central and Eastern Europe, IABSE Colloquium, Berlin, Germany, June 4–5, 1998.
11. Evaluation of Existing Steel and Composite Bridges, IABSE Workshop, Lausanne, Vol. 76 March 1997.
12. Extending the Lifespan of Structures, IABSE Symposium, San Francisco, CA, USA, August 23–25, 1995.
13. Structural Preservation of the Architectural Heritage: IABSE Symposium, Rome, Italy, September 15–17, 1993.
14. Remaining Structural Capacity, IABSE Colloquium, Copenhagen, Denmark, March 17–19, 1993
15. Length Effect on Fatigue of Wires and Strands: IABSE Workshop, Madrid, Spain, September 23–25, 1992

(III) IABSE SED Documents:

http://www.iabse.org/publications/onlineshop/index.php

1. Design for Robustness, Franz Knoll and Thomas Vogel, SED 11, 2009.
2. Cable Vibrations in Cable-Stayed Bridges, Elsa de Sa Caetano, SED 9, 2007.
3. Introduction to Safety and Reliability of Structures, Jörg Schneider, SED 5, 2nd edition, 2006.
4. Use of Fibre Reinforced Polymers in Bridge Construction, Thomas Keller, SED 7, 2003.
5. Vibrations in Structures—Induced by Man and Machines, H. Bachmann, W. Ammann, SED 3, 1987.

(IV) IABSE E-Learning: (Audio Visual presentations)

Refer to: http://www.iabse.org/

1. Lecture [4]: Wind-Induced Vibrations of Structures and Their Control, by Prof. M. ITO.
2. Lecture [10]: Life Cycle Management of Infrastructures: Towards and Integrated Approach of Design, Execution and Maintenance, Prof. A. van der Horst, (Keynote, 2007 Symposium, Weimar).
3. Lecture [13]: Retrofitting of Fatigue Damaged Steel Bridge Structures, Prof. Chitoshi Miki, Takuyo Konishi (Keynote, 2007 Symposium, Weimar).
4. Lecture [23]: Cable Vibrations in Cable-Stayed Bridges, Part (1): Assessment, by Dr. E. de S Caetano.

Appendix

B

List of Some Codes, Guidelines, Manuals, Documents, and Books on Assessment, Conservation, Evaluation, Inspection, Maintenance, Preservation, Rehabilitation, Repair, Retrofit, Strengthening and Upgrading Structural Performance

This Appendix includes a list of references on topics related to topics of SED 12: Repair, Rehabilitation, Retrofitting, Strengthening, Upgrading, . . . of Structures. The list of references in this Appendix is organized into the following sections: (I) Codes, Guidelines, Manuals, and Standards; (II) Books; (III) Documents, Bulletins, Reports, and Special publications published by International Associations Organizations; (V) Journals; (IV) Symposium, Workshop and Conferences; (VI) References on Historical and Heritage Structures; (VII) Videos and Presentations; (VIII) Relevant Websites and Relevant References; (IX) Checklists; (X) Terminology. For list of references on Diagnostic Crack Patterns and Causes of Deterioration in concrete structures, please refer to Appendix C in this document.

This list includes more than 600 references related to topics of SED 12. It is not a full or a complete list; however, it might be considered a wide-ranging list. The list is based on many searches on the Internet, however many other excellent references could be available which are not listed here.
(Note: Journal papers and Conference papers are not included in the lists in Appendix B.)

NOTE: Website links mentioned in this Appendix are active in January 2010. Website links might change without notice.

(I) Codes, Guidelines, Manuals, Standards (Partial List):

1. AASHTO CORE-1-M Guide for Commonly Recognized Structural Elements with 2002 and 2010 Interim Revisions, American Association of State Highway and Transportation Officials, Washington, D.C., USA.
2. AASHTO FIM-2-UL – Foundation Investigation Manual, 2nd Ed., USA, 1978.
3. AASHTO GBMS-1 – Guidelines for Bridge Management Systems, 1st Ed., USA, 1993.
4. AASHTO GMPC-2 – A Guide for Methods and Procedures in Contract Maintenance, 2nd Ed., 2002.
5. AASHTO IGSRB-1 – Inspectors' Guide for Shotcrete Repair of Bridges, USA, 1999.
6. AASHTO MBE-1-M – Manual for Bridge Evaluation, 1st Ed., with 2010 Interim Revisions, USA, 2008.

7. AASHTO MBI-1 – Movable Bridge Inspection, Evaluation, and Maintenance Manual, USA, 1998.
8. AASHTO MM-4 – Maintenance Manual for Roadways and Bridges, USA, 2007.
9. AASHTO MMS-1 – Guidelines for Maintenance Management Systems, 1st Ed., USA, 2004.
10. AASHTO RP-TANG-1 – Transportation Asset Management Guide, Prepared for NCHRP project 20-24(11), 2002.
11. AASHTO TF37-1 – Guide Specifications for Shotcrete Repair of Highway Bridges, USA, 1998.
12. AASHTO TF45-1 – Asset Management Data Collection Guide, AASHTO-AGC-ARTBA Task Force 45 Document, Washington, D.C., USA, 2006.
13. AASHTO-AGC-ARTBA TF32-1 – Manual for Corrosion Protection of Concrete Components in Bridges, USA, 1992.
14. ACI 562 – A new Concrete Repair Code, American Concrete Institute, will be in place by 2014 for incorporation by reference in IEBC. (ACI Committee 562, Evaluation, Repair, and Rehabilitation of Concrete Buildings).
15. ACMC – Asian Concrete Model Code, 2006, International Committee on Concrete Model Code for Asia, ICCMC, Level 2 Documents, Part III – Maintenance. Ch. 1: Scope; Ch. 2: Basis for Maintenance; Ch. 3: Inspection; Ch. 4: Deterioration Mechanism and Prediction; Ch. 5: Evaluation and Decision Making; Ch. 6: Remedial Action. (6.1) General, (6.2) Repairs, (6.3) Strengthening, (6.4) Other remedial measures; Ch. 7: Records.
16. Alberta Repair Manual For Concrete Bridge Elements, Technical Standards Branch, Alberta, Ver 02, Canada, 2005. http://www.transportation.alberta.ca/Content/docType30/Production/RpMConcBrEl2.pdf
17. Alberta Repair of Bridge Structural Steel Elements Manual, Technical Standards Branch, Alberta, Ver 01, Canada, 2004. http://www.transportation.alberta.ca/Content/docType30/Production/rprbrstrstel.pdf
18. AREMA – Bridge Inspection Handbook, American Railway Engineering Association, Lantham, Md, USA, 2008.
19. AS5100.7 – Bridge Design – Part 7: Rating of Existing Structures, Standards Australia, New South Wales, 2004.
20. ASCE/SEI 11-99 – Guideline for Structural Condition Assessment of Existing Buildings, 2000.
21. ASCE/SEI 41-06 – Seismic Rehabilitation of Existing Buildings, 2006. This new national consensus standard was developed from the FEMA 356.
22. ASCE 31-03 – Seismic Evaluation of Existing Buildings, American Society of Civil Engineers, 2002.
23. ASTM D4788-03 – Standard Test Method for Detecting Delaminations in Bridge Decks Using Infrared Thermography, American Society for Testing and Materials, USA, 2007.
24. ASTM WK 22503 – New Test Method for Determining Tensile Properties of Fiber Reinforced Polymer Matrix Composites Used for Strengthening of Civil Structures, 2009.
25. ASTM WK22346 – Determining Overlap Splice Tensile Properties of Fiber Reinforced Polymer Matrix Composites Used For Strengthening of Civil Structures, 2009.
26. Bridge Repair Manual of Finnish Road Administration (Silko) J. Lämsä, Bridge Engineering, Finnish Road Administration, Finland, 2003.
27. Bridge Repair Manual, RC 4300, Australian Rail Track Corporation, Issue A, Rev 1, 2006. http://extranet.artc.com.au/docs/engineering/tech_bulletins/manuals/section09/rc4300_bridge_repair_manual.pdf
28. BS HB 10141 – Handbook of Buildings: Service Life Planning. Part 1. General Principles. British Standards Institution, London, UK.1997.
29. Caltrans – Maintenance Manual Vol. 1, California Department of Transportation, USA, 2006. Includes sections on bridges, tunnels, Storm drainage, pavement, roads, ... http://www.dot.ca.gov/hq/maint/manual/maintman.htm
30. CSA A864-00 (R2005) – Guide to the Evaluation and Management of Concrete Structures Affected by Alkali-Aggregate Reaction, Canadian Standards Association, Canada, 2005.
31. CSA S448.1-93 – Repair of Reinforced Concrete in Buildings, Canadian Code, Canada, 2005.
32. CSA S478-95 (R2007) – Guideline on Durability in Buildings, Canada, 1995.
33. CSA S6-06 – Canadian Highway Bridge Design Code, 10th Ed., Sec. 14: Evaluation, Sec. 15: Rehabilitation and Repair, Canada, 2006.

34. DMRB – Design Manual for Roads and Bridges. The Stationery Office, Highways Agency, London, UK. http://www.standardsforhighways.co.uk/dmrb/
- Volume 3: Highway Structures: Inspection and Maintenance: Sec. 1: Inspection, Sec. 2: Maintenance, Sec. 3: Repair, Sec. 4: Assessment.
BA 16/97 The Assessment of Highway Bridges and Structures.
BA 30/94 Strengthening of Concrete Highway Structures Using Externally Bonded Plates.
BA 34/90 Technical Requirements for the Assessment and Strengthening Programme for Highway Structures.
BA 35/90 Inspection and Repair of Concrete Highway Structures.
BA 38/93 Assessment of the Fatigue Life of Corroded or Damaged Reinforcing Bars.
BA 39/93 Assessment of Reinforced Concrete Half-joints.
BA 43/94 Strengthening, Repair and Monitoring of Post-tensioned Concrete Bridge Decks.
BA 44/96 The Assessment of Concrete Highway Bridges and Structures.
BA 50/93 Post-tensioned Concrete Bridges. Planning, Organisation and Methods for Carrying Out Special Inspections.
BA 51/95 The Assessment of Concrete Structures Affected by Steel Corrosion.
BA 52/94 The Assessment of Concrete Structures Affected by Alkali Silica Reaction.
BA 54/94 Load Testing for Bridge Assessment.
BA 55/06 The Assessment of Bridge Substructures and Foundations, Retaining Walls and Buried Structures.
BA 72/03 Maintenance of Road Tunnels.
BA 74/06 Assessment of Scour at Highway Bridges.
BA 83/02 Cathodic Protection for Use in Reinforced Concrete Highway Structures.
BA 86/06 Advice Notes on the Non-Destructive Testing of Highway Structures.
BA 87/04 Management of Corrugated Steel Buried Structures.
BA 88/04 Management of Buried Concrete Box Structures.
BA 93/09 Structural Assessment of Bridges with Deck Hinges.
BD 21/01 The Assessment of Highway Bridges and Structures.
BD 27/86 Materials for the Repair of Concrete Highway Structures.
BD 34/90 Technical Requirements for the Assessment and Strengthening Programme for Highway Structures.
BD 44/95 The Assessment of Concrete Highway Bridges and Structures.
BD 45/93 Identification Marking of Highway Structures.
BD 46/92 Technical Requirements for the Assessment and Strengthening Programme for Highway Structures , Stage 2 – Modern Short Span Bridges.
BD 48/93 The Assessment and Strengthening of Highway Bridge Supports.
BD 50/92 Technical Requirements for the Assessment and Strengthening Programme for Highway Structures, Stage 3 – Long Span Bridges.
BD 53/95 Inspection and Records for Road Tunnels.
BD 54/93 Post-tensioned Concrete Bridges. Prioritisation of Special Inspections.
BD 56/10 The Assessment of Steel Highway Bridges and Structures.
BD 61/10 The Assessment of Composite Highway Bridges and Structures.
BD 62/07 As Built, Operational and Maintenance Records for Highway Structures.
BD 63/07 Inspection of Highway Structures.
BD 79/06 The Management of Sub-standard Highway Structures.
BD 81/02 Use of Compressive Membrane Action in Bridge Decks.
BD 86/07 The Assessment of Highway Bridges and Structures For The Effects of Special Types General, Order (STGO) and Special Order (SO) Vehicles.
BD 87/05 Maintenance Painting of Steelwork.
BD 89/03 The Conservation of Highway Structures.
- Volume 1: Highway Structures: Approval Procedures and General Design, Section 2: Other Procedural Documents.
BD 36/92 Evaluation of Maintenance Costs in Comparing Alternative Designs for Highway Structures.
BA 28/92 Evaluation of Maintenance Costs in Comparing Alternative Designs for Highway Structures.

BD 95/07 Treatment of Existing Structures on Highway Widening Schemes.
- Volume 7: Pavement Design and Maintenance; Section 3: Pavement Maintenance Assessment, Section 4: Pavement Maintenance Methods, Section 5 Surfacing and Surfacing Materials.

35. E-CFR, Title 23: Highways, Part 650: Bridges, Structures, Electronic Code of Federal Regulations, Federal Highway Administration, Washington, D.C., USA, 2010.
Subpart C: National Bridge Inspection Standards, Highways,
Subpart D: Highway Bridge Replacement and Rehabilitation Program.
http://ecfr.gpoaccess.gov/cgi/t/text/text-idx?c=ecfr&tpl=%2Findex.tpl
36. ECP 208-2005 – Egyptian Code of Practice For the Use of Fiber Reinforced Polymer (FRP) In the Construction Fields, Code No. ECP 208-2005.
37. EN 1504 – Products and systems for the protection and repair of concrete structures – Definitions – Requirements – Quality control and evaluation of conformity, European standard.
    – EN 1504 – Part 1: 2005 Definitions, Part 2: 2004 Surface protection systems for concrete,
    – Part 3:2005 Structural and non-structural repair, Part 4: 2004 Structural bonding,
    – Part 5:2004 Concrete injection, Part 6: 2006 Anchoring of reinforcing steel bar,
    – Part 7:2006 Reinforcement corrosion protection,
    – Part 8:2004 Quality control and evaluation of conformity,
    – Part 9:2008 General principles for use of products and systems,
    – Part 10:2003 Site application of products and systems and quality control of the works.
38. EN 1998-3: 2005 – Eurocode 8: Design of structures for earthquake resistance. Part 3: Assessment and retrofitting of buildings, European standard, 2005.
39. EN 1799 – Tests to measure the suitability of structural bonding agents for application to concrete surface, European standard, 1998.
40. EN 12190 – Determination of compressive strength of repair mortar, 1998.
41. EN 12617-2 – Shrinkage of crack injection product formulated with polymer binder – Part 2: Volumetric shrinkage, European standard, 2003.
42. EN 12617-3 – Determination of early age linear shrinkage for structural bonding agent, 2002.
43. EN 12636 – Determination of adhesion concrete to concrete, 1999.
44. EN 13395-3 – Determination of workability – Part 3: Test for flow of repair concrete, 2002.
45. EN 13395-4 – Determination of workability – Part 4: Application of repair mortar overhead, 2002.
46. EN 13733 – Determination of the durability of structural bonding agents, European standard, 2001.
47. EN 13791 – Assessment of in-situ compressive strength in structures and precast concrete components. Brussels: CEN, 2007.
48. *fib* – Model Code 2010, Chapter on Conservation, International Federation for Structural Concrete (*fib* – fédération internationale du béton), (Under Preparation).
49. GovHK – Building Maintenance Guidebook, Buildings Department, The Government of the Hong Kong SAR, 2002. http://www.bd.gov.hk/english/documents/code/bmg/BDG_ENG.pdf
50. GovHK – Interim Technical Guidelines on the Inspection, Assessment & Repair of Buildings for the Building Safety Inspection Scheme, Buildings Dept., Government of Hong Kong SAR, 1998.
51. IBC – International Building Code, by International Code Council, INC., Ch. 34: Existing Structures, USA, 2009.
52. ICC – Residential Seismic Rehabilitation, International Code Council, Code Master, 2009.
53. ICC FEBC 2007 – Florida Existing Building Code: Code & Commentary, ICC, International Code Council, Country Club Hills, 2007.
54. ICC IEBC 2006 – Appendix A: Code and Commentary for Seismic Retrofit of Existing Buildings, International Code Council, Country Club Hills, 2006.
55. ICC IEBC 2009 – International Existing Building Code, International Code Council, USA, 2009. Documents Contains requirements intended to encourage the uses and reuse of existing buildings. The scope covers repair, alteration, addition and change of occupancy for existing buildings and historic buildings, while achieving appropriate levels of safety without requiring full compliance with the new construction requirements in the building code.
56. ICC IPMC 2006 – International Property Maintenance Code, ICC, International Code Council, 2006.
57. ICC MREBC 2006 – Michigan Rehabilitation of Existing Building Code, ICC, 2006.
58. ICC VMV 2006 – Virginia Maintenance Code, ICC, International Code Council, 2006.
59. IITK-GSDMA – Guidelines: Seismic Evaluation and Strengthening of Buildings, Provisions with Commentary & Examples by D. C. Rai, Indian Institute of Technology Kanpur, India, 2005.

60. IITK-GSDMA EQ07-V1.0 – Review of Documents on Seismic Strengthening of Existing Buildings, Final Report: A – Earthquake Codes, by D.C. Rai, Project on Building Codes, India, 2001.
61. ISO 13822: 2001 – Bases for Design of Structures – Assessment of Existing Structures, International Organization for Standardization, 2001.
62. ISO 13823: 2008 – General principles on the design of structures for durability, 2008.
63. ISO TC 71/SC 7 – Maintenance and repair of concrete structures (Standards under development). ISO's umbrella code for maintenance and repair of concrete structures.
Part 1: General Principles for Maintenance and Repair; Part 2: Condition Assessment, and Design & Execution of Repairs; Part 3: Leakage due to Cracks; Part 4: Assessment and Repair of Earthquake Damage.
64. ISO WD TR 16475 – Guidelines for the Repair of Water-Leakage Cracks in Concrete Structures, Standard under development, ISO TC 71/SC 7 (Under development).
65. JBDPA – Seismic retrofit design and construction guidelines for existing reinforced concrete and steel encased reinforced concrete buildings using continuous fibre reinforcing materials, Japan Building Disaster Prevention Association, Tokyo, Japan, 1999.
66. JBDPA – Standard for evaluation of seismic capacity and guidelines for seismic retrofit design of existing reinforced concrete buildings, Japan Building Disaster Prevention Association, Tokyo, Japan, 1977, (revised 1990).
67. JSCE Guidelines for Concrete No. 4 – Standard Specifications For Concrete Structures – "Maintenance", 2nd Ed. 2007. 1st Ed. 2001, English ver. 2003. Japan Society of Civil Engineers, Japan. Consists of two parts: Part 1: Maintenance, Part 2: Standards for Maintenance.
Part 1 introduces a complete process from inspection to evaluation and judgment (including deterioration prediction) for concrete structure diagnosis, as well as matters related to the establishment of a maintenance plan. Part 2, includes updates of recent development related to each deterioration cause, and specifications for the maintenance of deteriorated structure, & for prevention of the deterioration. Additionally, fundamental introduction on the maintenance related to seismic strengthening of concrete structures is also provided.
68. JSCE Guidelines for Concrete No. 11 – Standards on Test Methods of Repairing Materials Used For Concrete Structures, Japan Society of Civil Engineers, Japan, 2007.
69. JSCE Guidelines for Retrofit of Concrete Structures – Draft, Concrete Library International, No. 36, December 2000, Japan. Ch. 7: Selection of Retrofitting Method. http://www.jsce.or.jp/committee/concrete/e/newsletter/newsletter01/recommendation/FRP-sheet/3.pdf
70. JSCE Recommendations for upgrading of concrete structures with use of continuous fiber sheets, Research Committee on Upgrading of Concrete Structures with Use of Continuous Fiber Sheet (2001), Concrete Engineering Series, No.41, JSCE, Tokyo.
71. JSCE – E 541 to E549 – Nine test methods are proposed to determine the properties of continuous fiber sheets for upgrading concrete structures, JSCE, Japan, 2000.
e.g.: E 547 – Test method for accelerated artificial exposure of continuous fiber sheets, 2000. http://www.jsce.or.jp/committee/concrete/e/newsletter/newsletter01/recommendation/FRP-sheet/2-8.pdf
72. JCI – Practical Guideline for Investigation, Repair and Strengthening of Cracked Concrete Structures, Japan Concrete Institute, Japan, 2003. http://www.jci-net.or.jp/e/publish/others/index.html
73. Manual of Maintenance of Steel Bridge Structures: Planning, Design, & construction for Maintenance and Durability, by Investigation Research Committee on Durability of Steel Bridges, Japan, 1993.
74. MBRC- COMAR-05-16-01 – Maryland Building Rehabilitation Code, DHCD, USA. 2007.
75. NJDOT – Design Manual for Bridges and Structures, State of New Jersey, New Jersey Department of Transportation (NJDOT), USA, 2005.
http://www.state.nj.us/transportation/eng/documents/BDMM/
Sect. 9A – Reconstruction and Rehabilitation Projects; Sect. 9B – Bridge Deck Rehabilitation; Sect. 9C – Bridge Deck Evaluation Survey and Guidelines for Restoration Work; Sect.12 – Bridge Deck Repair Design Guidelines; Sect. 45 – Seismic Design and Retrofit.
76. Ohio DOT – On-line Bridge Maintenance Manual Preventive Maintenance/ Repair Guidelines for Bridges and Culverts, Ohio Department of Transportation, USA.

Topics: Abutments, Approach Slabs, Arch Bridges, Backwalls, Beams, Bearings, Concrete Beams, Culverts, Decks, Drain Pipes Below Deck, Expansion Joints, Over-the-Side Drainage, Piers, Railings, Scuppers, Sidewalks, Slab Bridges, Steel Beams, Streambeds, Wearing Surface. Website:http://www.dot.state.oh.us/Divisions/HighwayOps/Structures/bridge%20operations%20and%20maintenance/PreventiveMaintenanceManual/Pages/default.aspx

77. Ontario Structure Rehabilitation Manual, Policy, Planning and Standards Division Engineering Standards Branch Bridge Office, Ministry of Transportation, Queen's Printer for Ontario, Canada, 2004. Part (1): Condition Surveys; Part (2): Rehabilitation Selection; Part (3): Contract Preparation; Part (4): Construction.
78. SETRA – France – Service d'Etudes Techniques des Routes et Autoroutes, Service d'études sur les transports, les routes et leurs aménagements.
    - Management Methods for Structural Works & Bridges, (Méthodes de gestion des Ouvrages d'Art (OA)) http://www.piles.setra.equipement.gouv.fr/rubrique.php3?id_rubrique=26
    - IQOA, Guides for Visual Inspection of Bridges. http://www.piles.setra. developpement-durable.gouv.fr/rubrique.php3?id_rubrique=204 http://www.piles.setra.developpementdurable.gouv.fr/rubrique.php3?id_rubrique=33
    - IQOA, Support information & presentations for visual inspection, http://www.piles. setra.developpement-durable.gouv.fr/article.php3?id_article=367
    - Statistics on Bridges. http://www.piles.setra.developpement-durable.gouv.fr/rubrique.php3?id_rubrique=254
    - IPD – L'Inspection Détaillée Périodique (Detailed Periodic Inspection) http://www.piles.setra.developpement-durable.gouv.fr/rubrique.php3?id_rubrique=205
    - Management of the Civil works: for the use of the departments and other local communities, Methodology guide, SETRA, May 2006 (In French: Gestion des ouvrages d'art: A l'usage des départements et autres collectivités locales, Guide méthodologique, SETRA), 2006. http://www.setra.equipement.gouv.fr/Gestion-des-ouvrages-d-art-A-l.html
    - DCE, Documents on Management & Repair of Bridges & Structures, Service d'études sur les transports, les routes et leurs aménagements, (Sétra). http://www.piles.setra.developpement-durable.gouv.fr/rubrique.php3?id_rubrique=177
    - Repair and reinforcement in reinforced and prestressed concrete Structure, CCTP, version V3.1, 12 November 2009. (In French: Reparations et Renforcement des Ouvrages en Beton Armé Et Précontraint). http://www.piles.setra.equipement.gouv.fr/IMG/doc/CCTP_BA-BP_V3-1_cle2eec5c.doc
    - Repair of Masonry Structures, CCTP Type, version 1.1 of 19 June 2008 (In French: Reparations des Ouvrages en Maçonnerie) http://www.piles.setra.equipement.gouv.fr/IMG/doc/CCTP_Maconnerie_V1.1__cle6c34c7.doc
    - Documents are available also for Repair of other structure types.
79. SETRA – France – Steel Concrete Composite Bridges – Sustainable Design Guide, by D. De Matteis, G. Chauvel, N. Cordier, P. Corfdir, R. Leconte, D. Le Faucheur, R. Leglise, Sétra, France, 2010. http://www.setra.equipement.gouv.fr/IMG/pdf/US_1013w_Steel-Concrete_Composite_Bridges_Sustainable_Design_Guide.pdf
80. SIA 269 – Code of Existing structures – Basis, (to be released in 2011), Switzerland.
81. Singapore – BCA Building and Construction Authority: Sustainable Construction Series,,2008. www.bca.gov.sg/sustainableconstruction/sustainable_series.html
    A Guide on Corrosion Protection for Steel Structures, BCA Sustainable Construction Series – 5, 2008. http://www.bca.gov.sg/SustainableConstruction/others/ sc_corrosion_book_final.pdf
82. Singapore BCA Building and Construction Authority – Periodic Structural Inspection http://www.bca.gov.sg/psi/ Building Owner's Guide (FAQs); Building Control Act (Part V) – Inspection of Buildings; Building Control (Inspection of Buildings) Regulations; Guidelines for Structural Engineers; Forms; Brochure: Safe Structures, Safe Buildings – Periodic Structural Inspection.
83. Singapore BCA Building and Construction Authority – Building Maintenance and Strata Management Act. http://www.bca.gov.sg/publications/BMSM/BMSM.html
84. Virginia USBC – Virginia Uniform Statewide Building Code, USA, 2006.
    USBC is divided into three stand-alone parts: Virginia Construction Code, 2006, (Part I). Virginia Rehabilitation Code, 2006, (Part II). Virginia Maintenance Code, 2006 (Part III).

(II) Books (Partial List):

85. Appraisal and Repair of Reinforced Concrete, by R. Holland, Thomas Telford, London, 1997.
86. Appraisal and Repair of Timber Structures, by P. Ross, Thomas Telford Publishing, 2002.
87. Assessment and Refurbishment of Steel Structures, by Z. Agocs, J. Brodniansky, J. Vican, J. Ziolko, Routledge, UK , 2005.
88. Bridge and Highway Structure Rehabilitation and Repair, by M. A. Khan, McGraw-Hill, 2010.
89. Bridge Decks: Construction, Rehabilitation, and Placement, by N. Bettigole, R. Robison, American Society of Civil Engineers (ASCE), 1997.
90. Bridge Engineering: Rehabilitation, and Maintenance of Modern Highway Bridges, by D. Tonias, J. Zhao, McGraw-Hill Professional; 2nd Ed., Oct. 2006.
91. Bridge Inspection and Rehabilitation. A Practical Guide, Edited by L. G. Silano, Parsons Brinckerhoff, Wiley-Interscience, 1993.
92. Bridge Maintenance, Inspection and Evaluation, by K. R. White, J. Minor, and K. N. Derucher, K. N., Marcel Dekker, Inc. USA, 1992.
93. Bridge Rehabilitation, by W. Radomski, Imperial College Press, UK, 2002.
94. Bridge Strengthening and Rehabilitation, by P. P. Xanthakos, Prentice Hall PTR, 1995.
95. Building Dilapidation and Rejuvenation in Hong Kong, editor by A. Y.T. Leung and C.Y. Yiu, City University of Hong Kong and the Hong Kong Institute of Surveyors, Hong Kong, 2004.
96. Building Facades: A Guide To Common Defects In Tropical Climates, by M. Y. L. Chew, L. H. Kang, & C. W. Wong, World Scientific Pub Co Inc., UK, 1998.
97. Building Failure, Practical Studies from the Building Research Establishment, by Building Research Establishment (BRE), The Construction Press, UK, 1978.
98. Building Failures: Diagnosis and Avoidance, by W. H. Ransom, 2nd Ed., Taylor & Francis, 1987.
99. Building Maintenance and Preservation: A Guide to Design and Management, by E. Mills, 2nd Ed., Butterworth–Heinemann, Oxford, UK, 1994.
100. Building Maintenance Management, B. Chanter & P. Swallow, Wiley-Blackwell, 2000.
101. Building Maintenance Technology in Tropical Climates, by C. Briffitt, NUS Press, 1995.
102. Building Pathology: Principles and Practice, by E. Watt, Blackwell Science, Oxford, UK, 1999.
103. Building Surveys and Reports, E. A. Noy & J. Douglas, Wiley-Blackwell, 2005.
104. Building Surveys, by P. Glover, Butterworth-Heinemann, UK, 2009.
105. Computer Aided Rehabilitation of Sewer and Storm Water Networks, by S. Saegrov, IWA Publishing, UK, 2006.
106. Concrete Bridge Practice: Construction Maintenance & Rehabilitation, by V. K. Raina, Shroff Publishers and Distributors Pvt. Ltd., 1st Ed., 2005.
107. Concrete Bridge Strengthening and Repair – Case Histories, by I.L. Kennedy-Reid, Thomas Telford, ICE, 2009.
108. Concrete Bridges: Inspection, Repair, Strengthening, Testing and Load Capacity Evaluation, by V. K. Raina, McGraw-Hill, 1994.
109. Concrete Construction Engineering Handbook, by E. G. Nawy, CRC Press, 2008.
110. Concrete Durability & Repair Technology (Creating With Concrete Series), by R. K. Dhir and M. J. McCarthy, Thomas Telford Publishing, 1999.
111. Concrete Problems: Causes and Cures, by J. C. Roke, McGraw-Hill Book, 1982.
112. Concrete Repair and Maintenance Illustrated: Problem Analysis, Repair Strategy, Techniques, by P. H. Emmons, R.S. Means Company, USA, 2001.
113. Concrete Structures: Materials, Maintenance, and Repair, by D. Campbell-Allen, and H. Roper, Harlow: Longman Scientific & Technical, USA, 1991.
114. Concrete Structures: Protection, Repair and Rehabilitation, by R. Dodge Woodson, Elsevier, 2009.
115. Concrete, S. Mindess and J. F. Young, Prentice-Hall, Inc., Englewood Cliffs, N.J. 07632, 1981.
116. Concrete: Building Pathology, by S. Macdonald, Wiley-Blackwell, 2002.
117. Conservation of Bridges: A Guide to Good Practice, by G. Tilly, Gifford & Partners, USA, 2002.
118. Construction Waterproofing Handbook by M. T. Kubal , McGraw-Hill Companies, 2008.
119. Corrosion & Corrosion Protection of Steel in Concrete, by E.d. Swamy, Sheffield Academic Press, Vol. I & II, UK, 1994.

120. Corrosion Damaged Concrete – Assessment and Repair, by P. Pullar-Strecker, Butterworth-Heinemann & CIRIA, UK, 1987.
121. Corrosion in Reinforced Concrete Structures, edited by H. Bühni, Woodhead Publishing, Cambridge, UK, 2005.
122. Corrosion of Steel in concrete, J. Broomfield, E & FN Spon, London, UK, 1997.
123. Corrosion of Steel in Concrete: Prevention, Diagnosis and Repair, by L. Bertolini, B. Elsener, P. Pedeferri, and R. B. Polder, Berlin: Wiley-VCH Verlag GmbH & Co. KGaA., Germany, 2004.
124. Corrosion of Steel in Concrete: Understanding, Investigation and Repair, by J. P. Broomfield, Taylor & Francis, 2nd Ed., UK, 2007.
125. Corrosion Protection of Steel Structures, by K.A. Chandler, and D. A. Bayliss, Oxford: Elsevier Science Publishers, 1985.
126. Corrosion-Damaged Concrete: Assessment and Repair, by P. Pullar-Strecker, Butterworth, UK, 1987.
127. Cracking and building movement, by P. Dickinson, N. Thornton, RICS Books, 2004.
128. Dampness in buildings, by A. Oliver, J. Douglas, J. S. Stirling, Wiley-Blackwell, 1997.
129. Dampness in buildings: diagnosis, treatment, instruments, by T. Alan Oxley, E. G. Gobert, Butterworth-Heinemann, UK, 1994.
130. Defects and Deterioration In Buildings, by B. A. Richardson, Construction Press, London, 2nd Ed., London and New York, 2001.
131. Defects in buildings, Great Britain, Property Services Agency, H.M.S.O., 1989.
132. Durability Design of Concrete Structures in Severe Environments, by O. E. Gjørv, Taylor & Francis, 2009.
133. Durability Design of Concrete Structures, by A. Sarja and E. Vesikari, Report of RILEM Technical Committee 130-CSL, E & FN SPON, 1996.
134. Durability of Concrete Structures and Constructions: Silos, Bunkers, Reservoirs, Water Towers, Retaining walls, by L. M. Poukhonto, Published, by A. A. Balkema, The Netherlands, 2003.
135. Durability of Concrete Structures: Investigation, Repair, Protection, by G. Mays, E & FN Spon, 1992.
136. Durability of Engineering Structures: Design, Repair and Maintenance, by J. Bijen, J., Woodhead Publishing Limited, UK, 2003.
137. Earthquake Resistant Concrete Structure, G. G. Penelis & A. J. Kappo, E & FN SPON – Chapman & Hall, London, UK, 1997.
138. Failure and Repair of Concrete Structures, by S. Champion, John Wiley & Sons Inc., 1961.
139. Failure Mechanisms in Building Construction, by David H. Nicastro, ASCE Press, 1997.
140. Failure, Distress and Repair of Concrete Structures, by N. Delatte, CRC Press, USA, 2009.
141. Fatigue of Structures and Materials, by J. Schijve, Kluwer Academic Publishers, 2001.
142. Foundation Failure, by C. Szechy, Concrete Publications Limited, 1958.
143. Foundation Repair Manual, by R. W. Brown, McGraw-hill Professional Publishing, 1999.
144. FRP Composites for Reinforced and Prestressed Concrete Structures, A Guide to Fundamentals and Design for Repair and Retrofit, by P. Balaguru, A. Nanni, J. Giancaspro, Spon Press, 2008.
145. FRP Strengthened RC Structures, by J G Teng, J F Chen, S T Smith, John Wiley & Sons, UK, 2002.
146. Fundamentals of Earthquake Engineering, by A. S. Elnashai, L. D. Sarno, John Wiley & Sons Ltd, 2008.
147. Guide to Concrete Repair, by G. Smoak, Books for Business, 2002.
148. Guidelines for Failure Investigation, by H. F. Greenspan, J. A. O'Kon, K. J. Beasley, J. S. Ward, ASCE, USA, 1989.
149. Handbook of Concrete Bridge Management, by F. A. Branco and J. B. Reston, ASCE Press, 2005.
150. Handbook on Nondestructive Testing of Concrete, V. M. Malhotra, N. J. Carino, CRC Press, 2004.
151. Handbook on Repairs and Rehabilitation of RCC Buildings, CPWD, Central Public Works Department (CPWD), Government of India, New Delhi, India, 2001.
152. Handbook on Seismic Retrofit of Buildings, by A. Chakrabarti, D. Menon, A. K. Sengupta, Narosa, India, 2008.
153. Home Renovation, by D. K. F. Ching, D. E. Miller, John Wiley & Sons, New York, USA, 1983.

154. Inspecting a House (For Pros By Pros), by R. Cauldwell, Taunton Press, USA, 2001.
155. Introduction to Structural Motion Control, by J. J. Connor, Prentice Hall Pearson Education, Inc., 2003.
156. Lessons from Failures of Concrete Structure, by J. Feld, ACI Monograph No. 1, USA, 1964.
157. Long Term Durability of Structural Materials, by P. J. M. Monteiro, K. P. Chong, J. Larsen-Basse, K. Komvopoulos, Elsevier Science; 1st Ed., 2001.
158. Maintenance Engineering Handbook, by L. R. Higgins, R. K. Mobley, R. Smith, 6th Ed., McGraw-Hill, USA, 2002.
159. Maintenance et Réparation des Ponts (Maintenance and repair of Bridges), by J. A. Calgaro, R. Lacroix, Presses de l'Ecole Nationale des Ponts et Chaussées, Paris, France, 1997.
160. Management of Bridges/Gestion des Ponts, Anglo-French liaison report, Rapport conjoint Franco-Britannique, Highways Agency (HA), Transportation Research Lab (TRL), Services d'Etudes Techniques des Routes et Autoroutes (SETRA), Laboratoire Central des Ponts et Chaussées (LCPC), published by Thomas Telford Ltd., 2005.
161. Mechanics of Materials Companion: Case Studies, Design, and Retrofit, by B. Yeigh, John Wiley and Sons, 2001.
162. Old waterfront walls: Management, maintenance, and rehabilitation, by R. N. Bray, P. F. B. Tatham, Taylor & Francis, 1992.
163. Parking Structures: Planning, Design, Construction, Maintenance and Repair, by A. P. Chrest, M. S. Smith, S. Bhuyan, Kluwer Academic Publishers, USA, 2001.
164. Pocket Surveying Buildings, by M. Hollis, RICS Books; 2 Rev Ed., 2006.
165. Recommendations for the inspection, maintenance and management of car park structures, ICE, Thomas Telford, 2002.
166. Refurbishment of Buildings and Bridges, by F. M. Mazzolani, M. Ivanyi, Springer, 2002.
167. Rehabilitation & Retrofitting of Structures, by KB Rajoria, A. Basa, MacMillan, India 2010.
168. Rehabilitation of Concrete Structures, by B. Vidivelli, Standard Publishers Distributors, India 2007.
169. Remedial Treatment of Buildings, by B. A. Richardson, Construction Press, London, 2nd Ed., Butterworth-Heinemann, Oxford, UK, 1995.
170. Renovation: A Complete Guide, by M. W. Litchfield, Sterling Publishing, USA, 1997.
171. Repair and Protection of Concrete Structures, by N.P. Mailvaganam, CRC Press, Boca Raton, 1992.
172. Repair and Renovation of Concrete Structures, Edited by M. R. Jones, & L. Zheng, R. Dhir, Thomas Telford, 2005.
173. Repair and Renovation of Modern Buildings, by I. Chandler, McGraw-Hill Inc., 1991.
174. Repair of Concrete Bridges, by G. P. Mallett, TRL, Thomas Telford Ltd. 1994.
175. Repair of Concrete Structures to EN 1504, Dansk Standard, Elsevier Butterworth-Heinemann, UK, 2004.
176. Repair, Protection and Waterproofing of Concrete Structures, by P. H. Perkins, Spon E & FN 3rd Ed., UK, 1997.
177. Retrofit Guide: Accessibility Modifications for Existing Buildings, by Barrier Free Environments Incorporated, UFAS, John Wiley and Sons, 1993.
178. Retrofitting Design of Building Structures, by Xilin Lu, CRC Press, 2010.
179. Seismic Assessment and Rehabilitation of Existing Buildings by S. T. Wasti, G. Zcebe, Springer-Verlag New York, 2010.
180. Seismic Design and Retrofit of Bridges, by M. J. N. Priestley, F. Seible, G. M. Calvi, Wiley-Interscience, 1996.
181. Seismic Design, Assessment and Retrofitting of Concrete Buildings: based on EN-Eurocode 8 (Geotechnical, Geological, and Earthquake Engineering), by M. N. Fardis, Springer 2009.
182. Seismic Retrofitting of RC Buildings: Computer Aided Seismic And Retrofitting Analysis Of Existing High Rise Reinforced Concrete Buildings, by R. R. Hussain, M.Wasim, VDM Verlag Dr. Muller Aktiengesellschaft & Co. KG, Pakistan 2010.
183. Seismic Risk Assessment and Retrofitting: With Special Emphasis on Existing Low Rise Structures (Geotechnical, Geological, and Earthquake Engineering), by A. Ilki, F. Karadogan, S. Pala, and E. Yuksel, Springer, Netherlands, 2009.

184. Service life estimation and extension of civil engineering structures, Edited by V M Karbhari, and L S Lee, CRC Press, (To be published). Part 1: Using Fibre Reinforced Polymer (FRP) Composites to Extend The Service-Life of Civil Engineering Structures, Part 2: Estimating the Service Life of Civil Engineering Structures.
185. Service Life Estimation of Civil Engineering Structures: Reinforced Concrete and FRP Rehabilitation Book Description, by S Lee L, L. S. Lee, V. M. Karbhari, CRC Press, 2010.
186. Steel Bridge Strengthening, by I. L. Reid, D. M. Milne, R. E. Craig, Thomas Telford, Case Studies. Highways Agency, UK, 2001.
187. Steel Corrosion in Concrete: Fundamentals and Civil Engineering Practice, by A. Bentur, S. Diamond, and N.S. BerkeLondon: E & FN Spon Bentur, A., Diamond, S. and Berke, N.S. 1997.
188. Strengthening and Rehabilitation of Civil Infrastructures Using Fibre-Reinforced Polymer (FRP) Composites, Edited, by L. C. Hollaway; J. G. Teng, CRC Press, 2008.
189. Strengthening of Reinforced Concrete Structures: Using Externally-Bonded FRP Composites in Structural and Civil Engineering by L. C. Hollaway, M. B. Leeming, Woodhead Publishing, Cambridge, England, 1999.
190. Structural Condition Assessment, by R. T. Ratay, John Wiley & Sons, 2005.
191. Structural defects reference manual for low-rise buildings, by M. F. Atkinson, Taylor & Francis, 2000.
192. Structural Renovation in Concrete, by Z. Li, C. Leung, Y. Xi, Taylor & Francis, 2009.
193. Structural Renovation of Buildings: Methods, Details, & Design Examples, by A. Newman, McGraw-Hill Professional Publishing, 2001.
194. Structural Surveying, by S. L. J. Mika & S. C. Desch, Macmillan Eduction, 2nd Ed., UK, 1988.
195. Subsidence damage to domestic buildings: A Guide to Good Technical Practice, by R. Driscoll & H. Skinner, BRE Trust, 2007.
196. Subsidence of Low-rise Buildings, The Institution of Structural Engineers: 2nd Ed., 2000.
197. Testing of concrete in structures, J. H. Bungey, S. G. Millard, and M. G. Grantham, Taylor and Francis, New York, USA, 2006.
198. The Repair of Concrete Structures, by R. T. L Allen, S.C. Edwards, & J. D. N. Shaw, 2nd Ed., Spon Press, 1993.
199. The Technology of Building Defects, by J. Hinks & G. Cook, Taylor & Francis, 1997.
200. Traditional Details: For Building Restoration, Renovation, and Rehabilitation, by C. G. Ramsey & H. R. Sleeper, John Wiley & Sons, April 1998.
201. Understanding Housing Defects, by D. Marshall, D. Worthing, R. Heath, Elsevier, 2003.
202. Underwater Concreting and Repair, by A.Mc Leish, Butterworth-Heinemann, UK, 1994.
203. Underwater Repair Technology, by J. H. Nixon, Gulf Professional Publishing, UK, 2000.
204. Upgrading and Retrofitting Water and Wastewater Treatment Plants, Water Environment Federation, WEF Press, USA, 2005.

205. Value Engineering: Practical Applications for Design, Construction, Maintenance & Operations, by A. J. Dell'Isola, R. S. Means Company, Inc., Kingston, MA, 1998.
206. 2010 National Renovation & Insurance Repair Estimator, by J. Russell, Craftsman Book Company, USA, 2010.
207. 2010 National Repair & Remodelling Estimator, by A. S. Paxton, Craftsman, USA, 2010.

Books Chapter (Partial List):

208. Bridge Engineering Handbook, edited, by W. F. Chen, CRC Press LLC, 2000.
Ch. 40: Seismic Retrofit Practice, by J. Roberts, B. Maroney.
Ch. 43: Seismic Retrofit Technology, by K. I. Keady, F. Alameddine, T. E. Sardo.
Ch. 49: Maintenance Inspection and Rating, by M. Vinayagamoorthy.
Ch. 50: Strengthening and Rehabilitation, by F. W. Klaiber, T. J. Wipf.
Ch. 53: Fatigue and Fracture, by R. J. Dexter and J. W. Fisher.

209. Bridge Management, by M. J. Ryall, Butterworth-Heinemann, UK, 2009.
Ch. 6: Repair, Ch. 7: Strengthening, Ch. 8: Bridge modification.
210. Earthquake Resistant Concrete Structure, by G.G. Penelis, and A.J. Kappos, E & FN Spon, 1997.
Ch. 13: Design & repair of strengthening, Ch. 14: technology of Shoring, repair and strengthening.
211. Evaluation and Rehabilitation, Ch. 15: from Prestressed Concrete Structures, by M. P. Collins & D. Mitchell, Canadian Prestressed Institute, p. 670–689, 1994.
212. The International Handbook of FRP Composites in Civil Engineering, editor Zoghi, CRC Press.
Ch. 5: Rehabilitation with NSM FRP Reinforcement, K. A. Soudki (under preparation).

(III) Documents, Bulletins, Technical Reports, and Special publications (Partial List):

213. ACI – Concrete Repair Manual – 3rd Ed., by ACI, ICRI, CS, BRE, NACE. American Concrete Institute, 2008.
214. ACI 228.2R-98 – Nondestructive test methods for evaluation of concrete structures, American Concrete Institute, Michigan, USA, 1998.
215. ACI 325.13R06 – Concrete Overlays for Pavement Rehabilitation, 2006.
216. ACI 341.3R-07 – Seismic Evaluation and Retrofit Techniques for Concrete Bridges, 2007.
217. ACI 364.1R-07 – Guide for Evaluation of Concrete Structures before Rehabilitation, American Concrete Institute, 2007.
218. ACI 437R-03 – Strength Evaluation of Existing Concrete Buildings, American Concrete Institute, 2003.
219. ACI 440.2R-08 – Guide for the Design and Construction of Externally Bonded FRP Systems for Strengthening Concrete Structures, 2008.
220. ACI 503.7-07 – Specification for Crack Repair by Epoxy Injection, USA, 2007.
221. ACI 546 R-04 – Concrete Repair Guide, Detroit, American Concrete Institute, 2004.
222. ACI 546.3R-06 – Guide for the Selection of Materials for the Repair of Concrete, 2006.
223. ACI C-10 – Repair and Rehabilitation of Concrete Structures, Case Studies Compilation from ACI Journals, 1990.
224. ACI C-20 – Repair and Rehabilitation II, Case Studies Compilation from ACI Journals, 1992.
225. ACI C-29 – Bridge Repair and Rehabilitation, 1995.
226. ACI E706 – Structural Crack Repair by Epoxy Injection, ACI RAP Bulletin 1, 2003.
227. ACI ICICRR – Repair and Rehabilitation: A Compilation from The Indian Concrete Journal, 2001.
228. ACI IPS-2 – Seismic Rehabilitation of Concrete Structures, 2007.
229. ACI SC27-93 – Repairing Concrete Bridges, American Concrete Institute, 1993.
230. ACI SCM-24 (07) – Concrete Repair Basics, 2007.
231. ACI SDC – Vision 2020: A Vision for the Concrete Repair, Protection, and Strengthening Industry, American Concrete Institute, Strategic Development Council (SDC), 2006. http://www.concretesdc.org/projects/Vision_2020.htm
232. ACI SP-258CD – Seismic Strengthening of Concrete Buildings Using FRP Composites, 2008.
233. ACRA HB 84-2006 – Guide to Concrete Repair and Protection, Australian Concrete Repair Association, ACRA, CSIRO, Sydney, Australia, 1996. http://www.acrassoc.com.au/Guide.htm
234. AISC 812-99 – Design Guide 12: Modification of Existing Steel Welded Moment Frame Connections for Seismic Resistance, American Institute of Steel Construction, 1999.
235. AISC 815-02 – Design Guide 15: Rehabilitation and Retrofit Guide: A Reference for Historic Shapes and Specifications, 2002.
236. API 510 – Pressure Vessel Inspection Code: In-Service Inspection, Rating, Repair, and Alteration, American Petroleum Institute, 2006.
237. API 570 – Piping Inspection Code: In-Service Inspection, Rating, Repair, and alteration of Piping Systems, 2009.
238. API Std 653 – Tank Inspection, Repair, Alteration, and Reconstruction, API Standard, 3rd, December 2001, Addendum 1, 2003.

239. ASA – Infrastructure Repair & Rehabilitation Using Shotcrete, American Shotcrete Association, Farmington Hills, MI, 2008. http://www.shotcrete.org/
240. ASCC – Specifications for Crack Repair – Position Statement No. 5, American Society of Concrete Contractors, 2003. http://www.ccagc.org/pdfs/ASCC_Position_Statement_05.pdf
241. ASCE Monograph 13 – Overcoming Barriers: Lifeline Seismic Improvement Programs, by C. Taylor, E. Mittler , L. Leval, and V. A. Reston, Technical Council on Lifeline Earthquake Engineering, (TCLEE),USA, 1998.
242. ASCE Monographs 15 – Guidelines for the Seismic Evaluation and Upgrade of Water Transmission Facilities, Editor: A. Eidinger, (TCLEE), USA, 1999.
243. ASCE MC475 – Existing Sewer Evaluation and Rehabilitation MOP FD-6, 3rd Ed., 2009.
244. ASHI – Standards of Practice and Code of Ethics, American Society of Home Inspectors, 2006. http://www.homeinspector.org/standards/default.aspx
245. ASTM STP 1352 – Water Problems in Building Exterior Walls: Evaluation, Prevention, and Repair, 1999.
246. ASTM STP 1444 – Building Facade Maintenance, Repair, and Inspection, American Society for Testing and Materials, 2004.
247. ASTM STP 1455 – Joining and Repair of Composites Structures, 2005.
248. ATC-40 – Seismic evaluation and retrofit of concrete buildings, Technology Council, State of California, Applied Technology Council, USA, 1996.
249. ATLSS 04-21 – Failure analysis of the U.S. 422 girder fracture, E. J. Kaufman, R. J. Connor, and J. W. Fisher, Final report, Center for Advanced Technology for Large Structural Systems, Lehigh Univ., Bethlehem, Pa., USA, 2004.
250. Australian Shotcrete Society – Recommended Practice Shotcreting in Australia, Concrete Institute of Australia, May 2008.
251. AWS D1-7-D1-7M-2010 – Guide for Strengthening and Repairing Existing Structures, American Welding Society, 2010.
252. AWWA – Condition Assessment of Norfolk's 36-Inch Raw Water Mains: Rehabilitation vs. Replacement, by G. Heisler, K. Turner, American Water Works Association, 2006.
253. AWWA – Rehabilitation of Water Transmission Mains: Two Case Histories, by R. L. Belcastro, American Water Works Association, 2001.
254. AWWA – Steel Water Storage Tanks: Design, Construction, Maintenance, and Repair, 2010.
255. BCA – The Diagnosis of Alkali-Silica Reaction, British Cement Association, 1992.
256. BRE – Effective cost analysis for repairing of corrosion damaged reinforced concrete structures: Residual life Models for Concrete Repair – Assessment of the Concrete Repair Process, Building Research Establishment Ltd, UK, 2003. http://projects.bre.co.uk/rebarcorrosioncost/
257. BRE 254 – Repair and Maintenance of Reinforced Concrete, by R. Currie and P. Robery, , UK, 1994.
258. BRE Digest 251 – Assessment of Damage in Low-rise Buildings, with Particular Reference to Progressive Foundation Movement, HMSO, UK, 1981.
259. BRE Digest 343 – Simple measuring & monitoring of movement in low-rise buildings, UK, 1989.
260. BRE Digest 353 – Damage to structures from ground-borne vibrations, HMSO, UK, 1990.
261. BRE Digest 366 – Structural Appraisal of Existing Buildings for Change of Use. Building Research Establishment, UK, 1991.
262. BRE Digest 434 – Corrosion of Reinforcement in Concrete – Electrochemical Monitoring, UK, 1998.
263. BRE Digest 444 – Part 3: Corrosion of Steel in Concrete – Protection and Remediation, UK, 2000.
264. BRE EP 77 – Achieving durable repaired concrete structures, Adopting a performance-based intervention strategy, by S. Matthews, M. Sarkkinen, and J. Morlidge, UK, 2007.
265. BRE EP 79 – Concrete repairs, Performance in service & current practice, by G. P. Tilly, & J. Jacobs, UK, 2007.
266. BRE FB 3 – Behaviour of concrete repair patches under propped and unpropped conditions: critical review of current knowledge and practices by T. D. G. Canisius and N. Waleed, UK, 2002.
267. BRE IP4/03 – Deterioration of cement-based building materials: lessons learnt, by A. M. Dunster and N. J. Crammond, UK, 2003.
268. BRE SD1 – Concrete in aggressive ground, Building Research Establishment, UK, 2005.

269. CBDG CCIP-024 – Guidance on the Assessment of Concrete Bridge, An updated and expanded report, by Concrete Bridge Development Group, Task Group and Members, TG 9, UK, 2007.
270. CBDG CCIP-036 – Enhancing the capacity of concrete bridges, Concrete Bridge Development Group, UK, 2008.
271. CCAA – Cement Concrete & Aggregates Australia. http://www.concrete.net.au/ http://www.concrete.net.au/concrete_structures/engineering.php (Refer to App. C for full list).
272. CEB-FIP 243 – Strategies for Testing and Assessment of Concrete Structures affected, by Reinforcement Corrosion, CEB – Bulletins d'Information, Euro-International Concrete Committee (CEB) & International Federation for Prestressing (FIP), 1998.
273. CIB Report: publication 331- CIB W080 -WG3 – State of the Art Report on Accelerated Laboratory Test Procedures and Correlation Between Laboratory Tests and Service Life Data, by B. Daniotti, F. Re Cecconi, (CIB: in French: Conseil International du Bâtiment, in English: International Council for Building, 2010). http://www.cibworld.nl/site/home/index.html
274. CIRIA 130 – Protection of Reinforced Concrete by Surface Treatments, Construction Industry Research and Information Association, UK, 1987.
275. CIRIA C519 – Action in the case of nonconformity of concrete structures, UK, 1999.
276. CIRIA C579 – Retention of masonry façades – best practice guide, by D. Lazarus, M. Bussell and P Ross, CIRIA, London, 2003.
277. CIRIA C589 – Retention of Masonry Façades – Best Practice Site Handbook, by D. Lazarus, M. Bussell and P. Ross, UK, 200.
278. CIRIA C595 – Strengthening metallic structures using externally-bonded fibre-reinforced-polymers, by J.M.C Cadei, T.J.Stratford, L.C. Hollaway, and W.G. Duckett, CIRIA, London, 2004.
279. CIRIA C656 – Masonry arch bridges: condition appraisal and remedial treatment, L. D. Mckibbins, C. Melbourne, N. Sawar, & C. S. Galliard, UK 2006.
280. CIRIA C664 – Iron and Steel Bridges: Condition appraisal and remedial treatment, G. P. Tilly, S. J. Matthews, D. Deacon, J. De Voy, & P. A. Jackson, UK 2008.
281. CPA – Standards for Repair and Protection of Concrete, British & European Standards, Corrosion Prevention Association, UK, 2005.
http://cpa.associationhouse.org.uk/default.php?cmd=210&doc_category=48
http://www.concrete.org.uk/services/fingertips_nuggets.asp?cmd=display&id=848
Factsheet (1) Introduction, Corrosion Prevention Association; (2) Surface Protection Systems for Concrete; European Standards for Repair & Protection of Concrete; (3) Structural & non-structural repair; Concrete Repair Standards News; (4) Structural bonding; List of BS for Repair and Protection of Concrete; (5) Concrete injection; Concrete Repair Standards News Issue 2; (8) Evaluation of Conformity; List of BS for Repair and Protection of Concrete; (9) General Principles for Protection & Repair; (10) Site Applications.
282. CPA – Technical Notes, by Corrosion Prevention Association.
http://www.corrosionprevention.org.uk/publications.html
No (1) Reinforced Concrete: History, Properties and Durability; (2) An Introduction to Electrochemical Rehabilitation Techniques; (3) Cathodic protection of steel in concrete – the international perspective; (4) Monitoring and maintenance of cathodic protection systems; (5) Corrosion mechanisms – an introduction to aqueous corrosion; (6) The principles and practice of galvanic cathodic protection for reinforced concrete structures; (7) Cathodic protection of early steel framed buildings; (8) Cathodic protection of steel in concrete – frequently asked questions; (9) Electrochemical Realkalisation of Steel Reinforced Concrete – A State of the Art Report; (10) Stray Current; (11) Impressed Current Anodes for the Cathodic Protection of Atmospherically Exposed Concrete; (12) Budget Cost and Anode Performance Information for Impressed Current Cathodic Protection of Reinforced Concrete Highway Bridges.
283. CS 137 – Guide to Evaluation and Repair of Concrete Structures in the Arabian Peninsula, Concrete Society, UK, 2002.
284. CS CPA – Repair Guidance Notes, Reproduced from Concrete, Concrete Society & Corrosion Prevention Association, 2008.
http://cpa.associationhouse.org.uk/default.php?cmd=210&doc_category=190

Note (1) Corrosion of steel in concrete; (2) Degradation of Concrete; (3) Fixing concrete; (4) Scope of EN1504 part 9; (5) En 1504 part2; (6) Mortars structural adhesives; (7) Concrete injection; (8) Anchoring of reinforcing steel; (9) Achieving successful repairs; (10) Performance based rehabilitation.

285. CS OP37 – Guide to maintenance, repair and monitoring of reinforced concrete structures, Concrete Society, UK, 2003.
286. CS TR22 – Non-Structural Cracks in Concrete, 3rd Ed., Concrete Society, UK, 1992.
287. CS TR26 – Repair of Concrete Damaged, by Reinforcement Corrosion, Technical Report, Concrete Society, UK, 1985.
288. CS TR32 – Analysis of Hardened Concrete, Concrete Society, UK, 1989.
289. CS TR33 – Assessment and Repair of Fire-Damaged Concrete Structures, Report of Concrete Society, 1999.
290. CS TR36 – Cathodic Protection of Reinforced Concrete, Concrete Society, UK, 1989.
291. CS TR37 – Model Specification for Cathodic Protection of Reinforced Concrete, Concrete Society, UK, 1991.
292. CS TR50 – Guide to Surface Treatments for Protection and Enhancement of Concrete, Concrete Society, UK, 1997.
293. CS TR54 – Diagnosis of deterioration in concrete structures, Identification of defects, evaluation and development of remedial action, Technical Report, Concrete Society, 2000.
294. CS TR55 – Design Guidance for Strengthening Concrete Structures using Fibre Composite Materials, Concrete Society, 2004.
295. CS TR57 – Strengthening Concrete Structures, Using Fibre Composite Materials; Acceptance, Inspection and Monitoring Concrete Society, Camberley, 2003.
296. CS TR67 – Movement, restraint and cracking in concrete structures, Concrete Society, UK, 2008.
297. CS TR69 – Repair of concrete structures with reference to BS EN 1504, Concrete Society, UK, 2009.
298. CTI TP07-08 – Corrosion-Induced Concrete Deterioration and Rehabilitation of Natural Draft Hyperbolic Cooling Tower, by L. Etcheverry, Corrosion Restoration Technologies, Cooling Technology Institute (CTI), 2007.
299. CTI TP99-10 – The Retrofit of a Large Cooling Tower at an Oil Refinery in Vienna in Combination with a Significant Noise Reduction, by H. van der Spek, V. S. Howden, Cooling Technology Institute (CTI), 1999.
300. Deterioration and Repair of above Ground Concrete Water Tanks in Ontario, Ontario Ministry of the Environment, Queen's Printer for Ontario, Canada, 1987.
301. DOCOMOMO – The Fair Face of Concrete: Conservation and Repair of Exposed Concrete, by W. De Jonge, and A. Doolaar, Preservation Dossier No. 2, DOCOMOMO International (International Working Party for Documentation and Conservation of Buildings, Sites and Neighborhoods of the Modern Movement), Eindhoven, 1997.
302. ECCS - JRC – Assessment of Existing Steel Structures: Recommendations for Estimation of Remaining Fatigue Life, by B. Kühn, M. Luki, A. Nussbaumer, H.-P. Günther, R. Helmerich, S. Herion, M.H. Kolstein, S. Walbridge, B. Androic, O. Dijkstra, Ö. Bucak, (Background documents on Eurocodes), (programme of CEN / TC 250), Editors: G. Sedlacek, F. Bijlaard, M. Géradin, A. Pinto and S. Dimova, 1st Ed., European Convention for Constructional Steelwork, 2008. http://www.steelconstruct.com/
303. ECCS 98 – Surface Protection Guide for Steelwork exposed to Atmosferic Environments, 1998.
304. ECCS 105 – Good Design Practice – a guideline for fatigue design, European Convention for Constructional Steelwork, 2000.
305. EFC 24 – Electrochemical Rehabilitation Methods for Reinforced Concrete Structures, by J Mietz, UK, European Federation of Corrosion, 1998. http://efcweb.org/Publications.html
306. FDOT 0510847 – Evaluation of Conventional Repair Techniques For Concrete Bridges, Summary of Final Report, SFR# 0510847, Florida Department of Transportation, 2000.
307. FDOT B352-09 – Project Planning Models For Florida's Bridge Management System, Summary of Final Report, 2004.
308. FDOT BB-879 – Development of Agency Maintenance Repair and Rehabilitation (MR&R) Cost Data for Florida's Bridge Management System, by J. O. Sobanjo, P. D. Thompson, State Maintenance Office Florida Department of Transportation, 2001.

309. FEMA – Design Manual for Retrofitting Flood-Prone Residential Structures, by Federal Emergency Management Agency, USA, University Press of the Pacific, 2004.
310. FEMA 154 – Rapid Visual Screening of Buildings for Potential Seismic Hazards: A Handbook. 2nd Ed., ATC-21, 2002.
311. FEMA 155 – Rapid Visual Screening of Buildings for Potential Seismic Hazards: Supporting Documentation, 2nd Ed., 2002.
312. FEMA 156 – Typical Costs for Seismic Rehabilitation of Existing Buildings. Volume 1: Summary, 2nd Ed., 1994.
313. FEMA 172 – NEHRP Handbook of Techniques for the Seismic Rehabilitation of Existing Buildings, 1992.
314. FEMA 306 – Evaluation of Earthquake Damaged Concrete & Masonry Wall Buildings (Basic Procedures Manual), ATC, 1998.
315. FEMA 307 – Evaluation of Earthquake Damaged Concrete and Masonry Wall Buildings: Technical Resources. Applied Technology Council, Redwood City, CA, 1998.
316. FEMA 308 – The Repair of Earthquake Damaged Concrete and Masonry Wall Buildings. Applied Technology Council. Redwood City, CA, 1999.
317. FEMA 310 – Handbook for the Seismic Evaluation of Buildings: A Prestandard. With ASCE, Washington, D.C., 1998.
318. FEMA 351 – Recommended Seismic Evaluation and Upgrade Criteria for Existing Welded Steel Moment-Frame Buildings. SAC Joint Venture Partnership. Washington, D.C., 2000.
319. FEMA 356 – Prestandard and Commentary for the Seismic Rehabilitation of Buildings, 2000.
320. FEMA 357 – Global Topics Report on the Prestandard and Commentary for the Seismic Rehabilitation of Buildings, 2000.
321. FEMA 399 – Incremental Seismic Rehabilitation of Retail Buildings, Providing Protection to People and Buildings, 2004.
322. FEMA 526 – Earthquake Safety Checklist, FEMA & NEHRP, 2005.
323. FEMA 547 – Techniques for the Seismic Rehabilitation of Existing Buildings, 2007.
324. FEMA P-420 – Engineering Guideline for Incremental Seismic Rehabilitation, 2009.
325. FHWA – Highway and Rail Transit Tunnel Inspection Manual 2005. http://www.fhwa.dot.gov/bridge/tunnel/inspectman00.cfm
326. FHWA – Highway and Rail Transit Tunnel Maintenance & Rehabilitation Manual, Federal Highway Administration, US. Department of Transportation, 2005. http://www.fhwa.dot.gov/bridge/tunnel/maintman00.cfm
327. FHWA – Bridge programs NBI data, Federal Highway Administration, National Bridge Inventory, 2007. http://fhwa.dot.gov/bridge/nbi.htm
328. FHWA IP-86-2 – Culvert Inspection Manual, by J. D. Arnoult, Federal Highway Administration, 1986.
329. FHWA-NBIS – National Bridge Inspection Standards. http://www.fhwa.dot.gov/bridge/nbis.htm, http://frwebgate.access.gpo.gov/cgi-bin/getdoc.cgi?dbname=2004_register&docid=FHWA%E2%80%932001%E2%80%938954.pdf
330. FHWA NC-2006-39 – Value Engineering and Cost Effectiveness of Various Fiber Reinforced Polymer Repair Systems-Part II, by O. Rosenboom, A. Miller, C. Walter, and S. Rizkalla, North Carolina Department of Transportation (NCDOT), North Carolina State University (NCSU), Constructed Facilities Lab (CFL), Report: RD-07-01, USA, 2007.
331. FHWA NHI 03-001 & 002 – Bridge Inspector's Reference Manual, by T. W. Ryan, P. E, R. A. Hartle, J. Eric Mann, L. J. Danovich, E. I. T., USA, October 2002, revised December 2006.
332. FHWA OH-2003/006 Environmental/Durability Evaluation of FRP Composite Strengthened Bridges, by N. F. Grace, Lawrence Technological University, Civil Engineering Department, Ohio Department of Transportation, USA, 2003.
333. FHWA PD-96-001 – Recording and coding guide for the structure inventory and appraisal of the nation's bridge, Federal Highway Administration, Washington, D.C., USA, 1996.
334. FHWA RD-01-02 – Reliability of visual inspection of highway bridges, Volume I: Final Report, M. E. Moore, B. M. Phares, B. A. Graybeal, D. D. Rolander, and G. A. Washer, USDOT, Washington, D.C., USA, 2001.

335. FHWA RD-03-031 – Distress Identification Manual for the Long-Term Pavement Performance Program, by J. S. Miller and W. Y. Bellingerm, (Fourth Revised Ed.), USA, June 2003. http://www.faa.gov/airports/great_lakes/airports_resources/certification_bulletins/media/09-07%20Attachment.pdf
336. FHWA RD-89-166 – A survey of localized cracking in steel bridges, C.E. Demers and J. W. Fisher, J. W, Vol. 1, Federal Highway Administration, U.S. Department of Transportation, USA, 1990.
337. FHWA RD-98-189 – Rehabilitation Of Prestressed Concrete Bridge Components by Non-Electrical (Conventional) Methods, by D. Whiting, H. Tabatabai, B. Stejskal, M. Nagi, USA, 1999.
338. FHWA RD-99-152 – Material and Procedures for Rapid Repair of Partial-Depth Spalls in Concrete Pavements, Manual of Practice, 1990.
339. *fib* Bulletin 14 – Technical Report on Externally Bonded FRP Reinforcement for RC Structures, International Federation for Structural Concrete, Lausanne, Switzerland, 2001. http://fib.epfl.ch/
340. *fib* Bulletin 17 – Management, Maintenance and Strengthening of Concrete Structures, Technical Report, 2002.
341. *fib* Bulletin 24 – Seismic assessment and retrofit of reinforced concrete buildings, State-of-art report, 2003.
342. *fib* Bulletin 35 – Retrofitting of concrete structures, by externally bonded FRPs, with emphasis on seismic applications Technical Report, 2006.
343. *fib* Bulletin 39 – Seismic design and retrofit – structural solutions, State-of-art report, 2007.
344. *fib* Bulletin 40 – FRP Reinforcement in RC Structures, Federation Internationale du Beton, 2007.
345. FIP – Repair and strengthening of concrete structures, FIP guide to good practice, Federations International du Béton, Thomas Telford, 1991.
346. Guide to Prepare a Building Maintenance Manual, Building Surveying Division, by D.C.W. Ho, S.M. Lo, and C.Y. Yiu, Hong Kong Institute of Surveyors, HK, 2008.
347. Heron – Durability of concrete structures, Editors A. C. W. M. Vrouwenvelder, R.B. Polder, Vol. 52, No. 4, Delft, The Netherlands, 2007. http://heron.tudelft.nl/52-4/2007_4.html
348. Heron – Inspection and Maintenance Strategies, by A. C. W. M. Vrouwenvelder, Vol. 39, No. 2, Delft, The Netherlands, 1994.
349. Heron – Maintenance and Repair of Concrete Structure, by J. M. J. M. Bijen, Vol. 34, No. 2, Delft, The Netherlands, 1989. http://heron.tudelft.nl/34-2/1.pdf
350. HSE RR 175 – Field studies of the effectiveness of concrete repairs, Phase 1 Report: Desk study and literature review, by N. J. R. Baldwin, E S King, Mott MacDonald Ltd. for the Health and Safety Executive, Crown, UK, 2003.
351. HSE RR 184 – Field studies of the effectiveness of concrete repairs, Phase 3 Report: Inspection of sites, sampling and testing at selected repair sites, by N. J. R. Baldwin, Mott MacDonald Ltd. for the Health and Safety Executive, Crown, 2003.
352. HSE RR 185 – Field Studies of the Effectiveness of Concrete Repairs, Phase 3a Report: An investigation of the performance of repairs and cathodic protection systems at the Dartford West Tunnel, by N. J. R. Baldwin, Mott MacDonald Ltd. for the Health and Safety Executive, Crown, 2003.
353. HSE RR 186 – Field studies of the effectiveness of concrete repairs, Phase 4 Report: Analysis of the effectiveness of concrete repairs and project findings, by N. J. R. Baldwin, E S King, Mott MacDonald Ltd. for the Health and Safety Executive, Crown, 2003.
354. HUD – A National Survey of Rehabilitation Enforcement Practices, by University of Illinois, Department of Housing and Urban Development, USA, 2007. http://www.huduser.org/portal/publications/destech/rehabsur.html
355. HUD – Barriers to Rehabilitation of Affordable Housing: Volume 1: Findings and Analysis, by Center for Urban Policy Research, J. Edward Bloustein School of Planning and Public Policy, B. Listokin, D. Listokin, 2001. http://www.huduser.org/portal/publications/destech/brah.html
356. HUD – Best Practices for Effecting the Rehabilitation of Affordable Housing, by E. J. Bloustein, M. Camp, A. Cander, Center for Urban Policy Research, K. crossney, D. Downs, Z. Kushel, D. Listokin, I. Litwin, T. McManus, M. Uschak, May, 2006. http://www.huduser.org/publications/pdf/BarriersVol1_part1.pdf
357. HUD – Rehabbing Flooded Houses: A Guide for Builders and Contractors, by Steven Winter Associates, Inc., 2008. http://www.huduser.org/portal/publications/pdf/Rehab_FloodedHouses.pdf

358. HUD – Residential Rehabilitation Inspection Guideline, by the U.S. Department of Housing and Urban Development, Office of Policy Development and Research, by the National Institute of Building Sciences, Washington, D.C., under Contract C-OPC-21204, 2000.
359. HUD – Smart Codes in Your Community: A Guide to Building Rehabilitation, by Building Technology Inc., 2005. http://www.huduser.org/portal/publications/destech/smartcodes.html
360. HUD – The Rehab Guide 1-9, by Steven Winter Associates, Inc., Department of Housing and Urban Development, USA, 2005. http://www.huduser.org/portal/publications/destech/rehabgds.html Vol. 1: Foundations, 1997; Vol. 2: Exterior Walls, 1999; Vol. 3: Roofs, 1999; Vol. 4: Windows & Doors, 1999; Vol. 5: Partitions, Ceilings, Floors, and Stairs, 1999; Vol. 6: Kitchens & Baths; Vol. 7: Electrical/Electronics; Vol. 8: HVAC/ Plumbing; Vol. 9: Site Work, 2006.
361. HUD-NAHB – Innovative Rehabilitation Provisions: A Demonstration of the Nationally Applicable Recommended Rehabilitation Provisions, by NAHB Research Center Inc., 1999.
362. HUD-NAHB – Innovative Rehabilitation Technologies – A State of the Art Overview, by Building Technology Inc., NAHB Research Center Inc., 1996.
363. HUD-NAHB – Rehabilitation Data Needs: A Building Industry Forum, by Harvard University, NAHB Research Center Inc., 2002.
364. HUD-NAHB – Residential Remodeling and Universal Design: Making Homes More Comfortable and Accessible, Barrier Free Environments, Inc., NAHB Research Center Inc., 1996.
365. HUD-NAHB – The Status of Building Regulations for Housing Rehabilitation, by Building Technology Inc., NAHB Research Center Inc., 1995.
366. HUD-NARRP – Nationally Applicable Recommended Rehabilitation Provisions, by Building Technology Inc., Koffel Associates, Inc., Melvyn Green & Associates, Inc., NAHB Research Center Inc., 1997.
367. HUD-NIBS – Guideline on Fire Ratings of Archaic Materials and Assemblies, National Institute of Building Sciences, 2000.
368. HUD-NIBS – Residential Rehabilitation Inspection Guide, 2006.
369. IABSE SED – Refer to Appendix A in this document, Section No. III, for a list of IABSE SED (Structural Engineering Documents) related to the topics of SED12, International Association for Bridge and Structural Engineering.
370. IBHS – Is Your Home Protected from Earthquake Disaster? A Homeowner's Guide to Earthquake Retrofit, Institute for Business & Home Safety, Boston, USA, 1999. http://www.disastersafety.org/
371. IBHS – Is Your Home Protected? A Homeowner's Guide to Earthquake Retrofit, prepared by the Institute for Business & Home Safety, Boston, USA, 2008.
372. ICE – FRP composites – Life Extension and Strengthening of Metallic Structure, by S.S.J. Moy (ed), Institution of Civil Engineers, London, 2001.
373. ICE-CARE – The Conservation Accreditation Register for Engineers, Institution of Civil Engineers, Institution of Civil Engineers, UK, http://www.ice.org.uk/care
374. ICRI – Technical Guidelines, International Concrete Repair Institute, USA. http://www.icri.org/publications/guidelines.asp
375. IStructE – Guide to surveys and inspections of buildings and associated structures, Institution of Structural Engineers, Institution of Structural Engineers, UK, 2008.
376. IStructE – Subsidence of low rise buildings (2nd ed.), UK, 2000.
377. IStructE – The operation and maintenance of bridge access gantries and runways, 2nd ed., UK, 2007.
378. LCPC-Sétra – Selection and Application of Repair and Protection of Concrete Structures: A Technical Guide, by P. Rossi, F. J. Ulm, D. Chauvel, 1996 (In French : Choix et application des produits de réparation et de protection des ouvrages en béton, Guide technique), Le Laboratoire Central des Ponts et Chaussees (LCPC), and SETRA, 1996.
379. MAE 09-01 – Design and Assessment Models and Spectra for Repaired Reinforced Concrete Structures, by G. Thermou et al, MAE Center Report No. 09-01, University of Illinois, Urbana-Champaign, USA, Mid-America Earthquake Center, 2009. https://www.ideals.illinois.edu/handle/2142/12906
380. MCEER 06-SP05 – Seismic Retrofitting Guidelines for Complex Steel Truss Highway Bridges, by T. Ho, R. Donikian, T. Ingham, C. Seim and A. Pan, Multidisciplinary and the National Center for Earthquake Engineering Research, USA, 2006.

381. MCEER 08-0011–Seismic Evaluation and Rehabilitation of Critical Components of Electrical Power Systems, by S. Ersoy, B. Feizi, A. Ashrafi and M. Ala Saadeghvaziri, USA, 2008.
382. MCEER 08-0015/16 – Seismic Performance Evaluation of Water Supply Systems, by Yu Wang and Thomas D. O'Rouke, USA, 2008.
383. MCEER-08-SP02 – Seismic Retrofitting Manual for Highway Structures: Part 1 – Bridges, by I. G. Buckle, I. Friedland, J. Mander, G. Martin, R. Nutt and M. Power, 2006, USA. Part 2 – Retaining Structures, Slopes, Tunnels, Culverts and Roadways, by I. G. Buckle, I. Friedland, J. Mander, G. Martin, R. Nutt and M. Power, 2006.
previously, FHWA – 1994 A Seismic Retrofitting Manual for Highway Bridges, by I.G. Buckle (Lead Author), I.M. Friedland and J.D. Cooper, MCEER/NCREE Bulletin Articles: Research, Apr 94 Vol 8 No2b.
384. MDOT- CEE-RR–2008/02 – Identification of causes and development of strategies for relieving structural distress in bridge abutments, by R. Burgueño, and Z. Li, Research report for Michigan Department of Transportation, Dept. of Civil and Environmental Engineering, Michigan State Univ., East Lansing, Mich., 2008.
385. NACE – Pipeline Rehabilitation, NACE Press, National Association of Corrosion Engineers, 2001.
386. NACE 21044 – Maintenance and Rehabilitation Considerations for Corrosion Control of Atmospherically Exposed Existing Steel-Reinforced Concrete Structures, SP0390-2006, The National Association of Corrosion Engineers, Houston TX, USA, 2006.
387. NACE 6/SSPC-SP 13 – Surface Preparation of Concrete, National Association of Corrosion Engineers, USA, 2003.
388. NACE RP0105-2005 – Liquid-Epoxy Coatings for External Repair, Rehabilitation, and Weld Joints on Buried Steel Pipelines, 2005.
389. NACE RPO296-2004 – Guidelines for Detection, Repair, and Mitigation of Cracking of Existing Petroleum Refinery Pressure Vessels in Wet H2S Environments, 2004.
390. NACE SP0109-2009 – Field Application of Bonded Tape Coatings for External Repair, Rehabilitation, and Weld Joints on Buried Metal Pipelines, 2009.
391. NACE SP0187-2008 – Design Considerations for Corrosion Control of Reinforcing Steel in Concrete, (formerly RP0187), 2008.
392. NACE SP0308-2008 – Inspection Methods for Corrosion Evaluation of Conventionally Reinforced Concrete Structures, 2008.
393. NAWSRC – Standards of Practice of the NAWSRC, Standards and Ethics of the National Association of Waterproofing and Structural Repair Contractors, 2006. www.nawsrc.org
394. NCHRP 280 – Guidelines For Evaluation and Repair of Damaged Steel Bridge Members, TRB, Transportation Research Board, National Cooperative Highway Research Program, USA, 1985.
395. NCHRP 353 SYNTHESIS – Inspection and Maintenance of Bridge Stay Cable Systems, A Synthesis of Highway Practice, Consultant: H. Tabatabai, TRB, Washington, D.C., 2005.
396. NCHRP 375 SYNTHESIS – Bridge Inspection Practices, A Synthesis of Highway Practice, Consultant: G. Hearn, TRB, Washington, D.C., 2007.
397. NCHRP 407 – Rapid replacement of bridge decks, by M. K. Tadros, M. C. Baishya, United States. Federal Highway Administration, National Research Council, TRB, USA, 1998.
398. NCHRP 514 – Bonded Repair and Retrofit of Concrete Structures Using FRP Composites Recommended Construction Specifications and Process Control Manual, by A. Mirmiran, M. Shahawy, A. Nanni & V. Karbhari, TRB, Washington, D.C., 2004.
399. NCHRP 590 – Multi-Objective Optimization for Bridge Management Systems, by V. Patidar, S. Labi, K. C. Sinha, TRB, USA, 2007.
400. NCHRP 608 GASB 34 – Methods for Condition Assessment and Preservation, Parsons Brinckerhoff, E. P. Chait, and Cambridge Systematic, Inc. TRB, Washington, D.C., 2008.
401. NCHRP 609 – Recommended Construction Specifications and Process Control Manual for Repair and Retrofit of Concrete Structures Using Bonded FRP Composites, by A. Mirmira, M. Shahawy, A. Nanni, V. Karbhari, B. Yalim, TRB, Washington, D.C., USA, 2008.
402. NCHRP 654 – Evaluation and Repair Procedures for Precast/Prestressed Concrete Girders with Longitudinal Cracking in the Web, by M. K. Tadros, S. S. Badie, C. Y. Tuan, TRB, National Cooperative Highway Research Program, Washington, D.C., USA, 2010.

403. NCHRP Synthesis 234 – Settlement of Bridge Approaches (The Bump at the End of the Bridge), by J. L. Briaud, R.W. James, and S.B. Hoffman, Transportation Research Board, National Research Council, Washington D.C., USA, 1997.
404. NCHRP Synthesis 303 – Assessment and Rehabilitation of Existing Culverts,by Consultant: D. C. Wyant, TRB, Washington, D.C., USA, 2002.
405. NCHRP Synthesis 389 – Performance-Based Contracting for Maintenance, Consultant W. A. Hyman, TRB, Washington, D.C., USA, 2009.
406. NCHRP Web Document 035 – Rehabilitation Strategies for Highway Pavements (Project C1-38), by K. T. Hall, C. E. Correa, S. H. Carpenter, R. P. Elliot, 2001.
407. NCHRP Web Document 050 – Repair and Rehabilitation of Bridge Components Containing Epoxy-Coated Reinforcement, by A. A. Sohanghpurwala, W. T. Scannell, W. H. Hartt, (Project D10-37C), TRB, 2002.
408. NCHRP Web Document 110 – Supplementary Cementitious Materials to Enhance Durability of Concrete Bridge Decks, J. S. Lawler, J. D. Connolly, P. D. Krauss, S. L. Tracy, B. E. Ankenman, and NCHRP Project 18-08A, 2005.
409. NCPP – National Center for Pavement Preservation; Concrete Pavement Preservation Publications: Diamond Grinding, Dowel Bar Retrofit, Rehabilitation/Repair, Noise. http://www.pavementpreservation.org/concrete/
410. NCPP – Pavement Preservation Checklist Series, FHWA, Federal Highway Administration (10 Documents). http://www.fhwa.dot.gov/pavement/preservation/ppcl00.cfm
411. NICEE – Guidelines for earthquake resistant non-engineered construction, by (IAEE) International Association of Earthquake Engineering; (NICEE) National Information Center of Earthquake Engineering, 2004, India & Japan. http://www.nicee.org/IAEE_English.php
412. NRC-CNRC-Construction Technology Update No. 59 – Repairs to Restore Serviceability in Concrete Structures, by N. P. Mailvaganam and L. Mitchell, National Research Council, Canada, 2003. http://www.nrc-cnrc.gc.ca/obj/irc/doc/ctu-n59_eng.pdf
413. NRC-CNRC-NRCC-46624 – Durability of concrete structures: prevention, evaluation, inspection, repair and prediction, by D. Cusson, B. Isgor, 2004.
www.nrc-cnrc.gc.ca/obj/irc/doc/pubs/nrcc46624/nrcc46624.pdf
414. NRMCA – National Ready Mixed Concrete Association. http://www.nrmca.org/
(Refer to App. C for full list).
415. NTIS – Evaluation, Maintenance, and Rehabilitation. Overlays on Horizontal Concrete Surfaces, by R. L. Campbell, Technical Report REMR-CS-42, National Technical Information Service, 1994.
416. OECD – Bridge Inspection, A report prepared, by Organisation for Economic Co-operation and Development, Road Research Group, France, 1976.
417. OECD – Bridge Management, Organisation for Economic Co-operation and development, 1992.
418. OECD – Bridge rehabilitation and strengthening, by Organization for Economic Co-operation and Development, Paris, 1983.
419. OECD – Repairing Bridge Substructures, by Organization for Economic Co-operation and Development, Paris, 1995.
420. PCA ACPA – Concrete Pavement Rehabilitation Guide for Load Transfer Restoration, JP001P, Portland Cement Association & American Concrete Pavement Association, IL, USA, 1997.
421. PCA ACPA – Guidelines for Protection and Repair of Concrete Pavements Exposed to Rain During Construction, American Concrete Pavement Association, Arlington Heights, IL, USA, 1987.
422. PCA ACPA TB002P – Guidelines for Full-Depth Repair, IL, 1995.
423. PCA ACPA TB003P – Guidelines for Partial-Depth Repair, IL, 1998.
424. PCA ACPA TB012P – Joint and Crack Sealing and Repair for Concrete Pavements, IL, 1993.
425. PCA IS001–Effects of Substances on Concrete and Guide to Protective Treatment, Portland Cement Association, Skokie, IL, 2007.
426. PCA IS144 – Resurfacing Concrete Floors, Portland Cement Association, Skokie, IL, 1996.
427. PCA IS526 – Repair of Portland Cement Plaster (Stucco), Skokie, IL, 2001.
428. PCA LT318 – Interlocking Concrete Pavement Distress Manual Item, Portland Cement Association, Skokie, IL, 2008.
429. PCA PT315 – Identification of Alkali-Silica Reactivity in Highway Structures, IL, 2003.

430. PCA SN2197 – Investigation of Deteriorated Concrete in Pavements, by F. M. Miller, R. J. Detwiler, and L. J Powers, IL, 2000.
431. PCI – Maintenance manual for precast parking structures, by PCI Committee on Parking Structures, Precast/Prestressed Concrete Institute, USA, 2004. www.pci.org/pdf/publications/parking/MNL_136_04.pdf
432. PIARC – Good Practice for the Operation & Maintenance of Road Tunnels, World Road Association, France, 2005. http://publications.piarc.org/ressources/publications_files/1/625-05-13-VCD.pdf
433. PIARC – Study on Rehabilitation Actions on Concrete Bridges, Technical Committee on Road Bridges and other Structures (C11), World Road Association, 2005. http://publications.piarc.org/ressources/publications_files/1/629-11-15-VCD.pdf
434. PWRI – Ground Anchor Maintenance Management Manual, by Public Works Research Institute and Japan Anchor Association, Kajima Institute Publishing Co., Ltd, Japan, 2008.
435. PWRI-MCEER 88-JPML – Manual for Repair Methods of Civil Engineering Structures Damaged by Earthquakes, Public Works Research Institute, Ministry of Construction, Japan 1986.
436. REHABCON IPS-2000-00063 – Strategy for Maintenance and Rehabilitation in Concrete Structures, DG Enterprise of the European Commission, 2004.
437. Repair and Rehabilitation of Dams: Case studies, by J. E. McDonald, N. F. Curtis, Repair, Evaluation, Maintenance, and Rehabilitation Research Program, U.S. Corps of Engineers, Engineer Research and Development Center, 1999.
438. Report IS-06-02 – Design Guidelines for the Use of HM Strips: Strengthening of Steel Concrete Composite Bridges with High Modulus Carbon Fiber Reinforced Polymer (CFRP) Strips, by D. Schnerch, M. Dawood, and S. Rizkalla, Technical Report, North Carolina State University (NCSU), Constructed Facilities Lab (CFL), Raleigh, USA, 2006 (updated June 2007).
439. RILEM – Recommendations TC 139-DBS – Durability of Building Sealants Durability test method – Determination of changes in adhesion, cohesion and appearance of elastic weatherproofing sealants for high movement façade joints after exposure to artificial weathering, International Union of Laboratories and Experts in Construction Materials, System and Structures, 2001. www.rilem.net/
440. RILEM Report rep 021 – Durability of Building Sealants – State-of-the-Art Report of RILEM TC 139-DBS, Edited by A. T. Wolf, 1999.
441. RILEM Report rep 026 – Towards Sustainable Roofing – Report of CIB W.83/RILEM TC 166-RMS – Environmental Task Group, Edited by CIB W 83/RILEM TC 166-RMS, 2003.
442. RILEM Report rep 027 – Condition Assessment of Roofs – Final Report of CIB W.83/RILEM TC 166-RMS – Condition Assessment Task Group, Edited by CIB W 83/RILEM TC 166-RMS, 2003.
443. RILEM Report rep 028 – Characterisation of Old Mortars with Respect to their Repair – Final Report of RILEM TC 167-COM, Edited by C. Groot, G. Ashall and J. Hughes, 2004.
444. RILEM Report rep 029 – Pavement Performance Prediction and Evaluation (PPPE) – Report of RILEM TC 152-PBM and 182-PEB, Edited by M. Partl and H. Piber, 2001.
445. RILEM Report rep 032 – Probabilistic Assessment of Existing Structures – JCSS Report, Edited by D. Diamantidis, 2001.
446. RILEM Report rep 042 – Acoustic Emission and Related Non-destructive Evaluation Techniques for Crack Detection and Damage Evaluation in Concrete – Final Report of RILEM Technical Committee 212-ACD, Edited by M. Ohtsu, 2010.
447. SB-ICA – Guideline for Inspection and Condition Assessment of Railway Bridges. Prepared by Sustainable Bridges – a project within EU FP6, Sustainable Bridge, 2007. Available from: www.sustainablebridges.net.
448. SB-MON – Monitoring Guidelines for Railway Bridges. Prepared by Sustainable Bridges – a project within EU FP6, 2007. Available from: www.sustainablebridges.net
449. SB-STR – Repair and Strengthening of Railway Bridges – Guideline. Prepared by Sustainable Bridges – a project within EU FP6, 2007. Available from: www.sustainablebridges.net
450. SEAOC RETUB – Retrofit of Existing Tilt-Up Buildings & Infill Frame, Structural Engineers Association of California, 1996.
451. SEI/ASCE 11-99 – Guideline for Structural Condition Assessment of Existing Buildings, Structures Engineering Institute & American Society of Civil Engineers, 2000.

452. Seismic retrofitting of steel and composite building structures, by L. Di Sarno and A.S. Elnashai, Mid – America Earthquake Center Report, University of Illinois at Urbana – Champaign, IL, USA, 2002. http://www.ideals.illinois.edu/bitstream/handle/2142/9265/Report02-01.pdf
453. SSPC-PA Guide 5 – Guide to Maintenance Coating of Steel Structures in Atmospheric Service, Protective Coatings Specialist (PCS), Steel Structures Painting Council, 2002, http://www.sspc.org/standards/pascopes.html# PA5
454. SSPC Paint 37 – Waterborne Epoxy Coating for Cementitious Substrates, Performance-Based, 2004.
455. SSPC SP COM – Surface Preparation Commentary for Steel and Concrete Substrates, November, 2004.
456. The performance of concrete bridges: A survey of 200 highway bridges, by E. J. Wallbank, Report prepared for the DOT, HMSO, London, April 1989.
457. TIRC/NYSDO – Bridge element deterioration rates, by A. K. Agrawal, A. Kawaguchi, and Z. Chen, Project No. C-01-51, Transportation Infrastructure Research Consortium & New York State Department of Transportation, 2009.
458. TRB – SHRP-S-336 – Techniques for Concrete Removal and Bar Cleaning on Bridge Rehabilitation Projects, Transportation Research Board, Strategic Highway Research Program, National Research Council, Washington, DC, USA, 1992. http://www.trb.org/publications/shrp/SHRP-S-336.pdf
459. TRB – SHRP-S-347 – Chloride Removal Implementation Guide, by Bennett, J.E. & Schue, T.J., Report, Washington DC, 1993.
460. TRB – SHRP-S-360 – Concrete Bridge Protection, Repair, and Rehabilitation Relative to Reinforcement Corrosion: A Methods Application Manual, NRC, Washington DC, 1993.
461. TRB – SHRP-S-665–Concrete Bridge Protection & Rehabilitation: Chemical & Physical Techniques, NRC, Washington DC, 1993.
462. TRB – SHRP-S-666 – Concrete Bridge Protection & Rehabilitation: Corrosion Inhibitors & Polymers, NRC, Washington DC, 1993.
463. TRB – Active heating infrared thermography for detection of subsurface bridge deck deterioration, K. Maser, Fibnal Report for Highway IDEA Project 101, Transportation Research Board, Washington, D.C., USA, 2004.
464. TRB Circular E-C139 – Use of Accelerated Pavement Testing to Evaluate Maintenance and Pavement Preservation Treatments, Full-Scale and Accelerated Pavement Testing Committee, Transportation Research Board, USA, Sept. 2009.
465. TRIS – ITRC FR 95/96-1- Evaluation of bridge deck delamination investigation methods, H. L. dos Reis, and M. D. Baright, Transportation Research Information Services, Edwardsville, Ill, USA, 1999.
466. TRL – PPR053 – Strengthening of concrete structures using near surface mounted FRP reinforcement, by A F Daly, J. Shave, S. Denton, Transport Research Laboratory, UK, 2006.
467. TRL – RR078 – Corrosion of reinforcement: an assessment of twelve concrete bridges after 50 years service, by PR Vassie, UK, 1986.
468. TRL – RR109 – Durability of concrete repairs: the effect of steel cleaning procedures, by PR Vassie, UK, 1987.
469. TRL – RR150 – Repair of cracked reinforced concrete: assessment of corrosion protection, by AJJ Calder, Thompson DM, UK, 1988.
470. TRL – RR276 – Assessment of trough to crossbeam connections in orthotropic steel bridge decks, by C. Beales C, UK, 1990.
471. TRL 283 – The durability of bonded steel plates for strengthening concrete highway bridges – a 20 year trial, by AJJ Calder, UK, 1997.
472. TRL Contractor Report 284 – Assessment of repair and strengthening techniques for brick and stone masonry arch bridges, by D. Ashurst, UK, 1992.
473. UN-HABITAT CHS-OP-82-1 – The rehabilitation of existing housing stock, Report of the expert group meeting on human settlements management, with special reference to the rehabilitation of existing housing stock, United Nations Human Settlements Programme, New Delhi, 1982. www.unhabitat.org/downloads/docs/3618_15959_CHS-OP-82-13.pdf
474. UN-HABITAT HS/242/91E – Report of the International Workshop on Building Maintenance Strategy, 1991.

475. UN-HABITAT HS-079-85 E – Rehabilitation of inner-city areas: Feasible strategies, UNCHS (Habitat), 1985.
476. USDA – Deep Patch Road Embankment Repair Application Guide, by S. W. Musser, C. Denning, United States Department of Agriculture, Forest Service Technology & Development Program, October 2005. www.fs.fed.us/t-d/pubs/pdf/05771204.pdf
477. WHRP 02-003- Assessment and rehabilitation strategies/guidelines to maximize the service life of concrete structures, by T. Adams, J. Pincheira, and Y. H. Huang, Wisconsin Highway Research Program, Madison, Wis., 2002. http://www.whrp.org/research-areas/structures/structures_0092-00-17.html
478. WHRP 09-04 Evaluation of Select Methods of Corrosion Control, Corrosion Prevention, and Repair in Reinforced Concrete Bridges, by H. Tabatabai, Al Ghorbanpoor, M. D. Pritzl, Wisconsin Highway Research Program, USA, 2009.

(IV) Technical Journals (Partial List):

479. AASTHO TSP2 – Pavement Preservation Journal, American Association of State Highway and Transportation Officials, USA. http://www.tsp2.org/library/document/pavement-preservation-journal-winter-2009
480. ACI – ICRI – Concrete Repair Bulletin, American Concrete Institute, USA. http://www.concrete.org/members/periodicaldescriptions.htm#CRB http://www.icri.org/publications/bulletin.asp
481. ASCE – Journal of Performance of Constructed Facilities, American Society of Civil Engineers. http://www.asce.org/Journal.aspx?id=2147486724
482. CIB –International Journal of Disaster Resilience in the Built Environment, International Council for Building. http://www. world.nl/site/searchn/results.html?q=rehabilitation
483. CRA – Cracking Matters Journal, Concrete Repair Association. UK. http://www.concreterepair.org.uk/cracmatters.html
484. IABSE SEI –Refer to Appendix A in this document, Section No. I, for a list of Papers published in IABSE SEI Journal on topics related to SED 12: Rehabilitation, Repair, Restoration, Retrofit, Stiffening, Strengthening, Upgrading, ...
485. NTHP–Preservation Magazine, National Trust for Historic Preservation. http://www.preservationnation.org/magazine/inside-our-january-february-issue.html

Note: References mentioned above present a partial list of specialized Journals related of the topics of SED 12. Several other excellent Journals were also organized which might not be mentioned in this partial list. Moreover, many of the Journals on Structural Engineering topics include papers on Assessment, Inspections, Evaluation, Conservation, Maintenance, Preservation, Repair, Rehabilitation, Retrofitting, Strengthening, Upgrading, . . . of Structures.

(V) Symposium, Workshop and Conferences (Partial List):

486. Bridge Management 1: Inspection, Maintenance, Assessment and Repair, Proc. 1st International Conference on Bridge Management, Edited by J. E. Harding, G. A. R. Parke, M.J. Ryall, Elsevier, London, 1990.
487. Bridge Management 2: Inspection, Maintenance, Assessment and Repair, Proc. 2nd Inter. Edited by J. E. Harding, G. A. R. Parke, M. J. Ryall, Thomas Telford, London, 1993.
488. Bridge Management 3: Inspection, Maintenance, Assessment and Repair, Proc. 3rd Inter., Edited by J. E. Harding, G. A. R. Parke and M. J. Ryall, F&N Spon, London,1996.
489. Bridge Management 4: Inspection, Maintenance, Assessment and Repair, Proc. 4th Inter., Edited by M. J. Ryall, G. A. R. Parke and J. E. Harding, Thomas Telford, London, 2000.
490. Bridge Management 5: Inspection, Maintenance, Assessment and Repair, Proc. 5th Inter., Edited by G. A. R. Parke, P. Disney, Thomas Telford, London, 2005.

491. Bridge Modification: Conference Proceeding, Edited by B. Pritchard, Thomas Telford, March 1994.
492. Bridge Modification 2: Stronger and Safer Bridges, Edited by B. Pritchard, Thomas Telford Ltd., 1997.
493. Bridge Rehabilitation: 3rd International Workshop, Ernst & Sohn Pub., Darmstadt, Germany, 1992.
494. Building Pathology: 2nd International Symposium on Building Pathology, Durability & Rehabilitation – Learning from Errors and Defects in Buildings, CIB W086, Lisboa, Portugal, November 2003.
495. Concrete repair, rehabilitation and protection, edited by R. K. Dhir and M. R. Jones, Proceedings of the International Conference, University of Dundee, Scotland, UK, June 1996.
496. Concrete Solutions, 3rd International Conference on Concrete Repair, Padova University, Taylorand Francis Group, Italy, June 2009.
497. Eighth International Symposium on Fiber Reinforced Polymer Reinforcement for Concrete Structures, (FRPRCS-8), Chair: T. C. Triantafillou, University of Patras, Patras, Greece, July 2007.
498. Failures, Rehabilitation and Retrofitting of Bridges & Aqueducts, International Seminar, Bombay, India, 1994.
499. Fatigue design 2009, ECCS, Senlis, France, November 2009. http://www.steelconstruct.com/
500. Fifth International Conference on Bridge Maintenance, Safety and Management, TRB (IABMAS 2010), in Philadelphia, Pennsylvania, July 2010.
501. Fifth International Conference on Water Repellent Treatment of Building Materials, Aedificatio Publishers, Freiburg, Editors H. De Clereq, and A. E. Charola, Brussels, Belgium April 2008.
502. IABSE Conferences: Refer to Appendix A in this document, Section No. II, for a list of IABSE Conferences related to the topics of SED12.
503. ICCRRR-1 (2005) Concrete Repair, Rehabilitation and Retrofitting, Edited by M. G. Alexander, H. D. Beushausen, F. Dehn, P. Moyo, published by Taylor & Francis Group, 2006.
504. ICCRRR-2 (2008) Concrete Repair, Rehabilitation and Retrofitting, 2nd International Conf., Edited by M. G. Alexander, H. D. Beushausen, F. Dehn, P. Moyo, CRC Press, 2009.
505. Innovative systems for seismic repair and rehabilitation of structures : design and application, Edited by A. S. Mosallan, proceedings of Second Conference on Seismic Repair and Rehabilitation of Structures (SRRS2), Fullerton, California, March 2000.
506. Intelligent Building Maintenance – A Novel Discipline, by E. CY Yiu, ISSF Conference 2008, Hong Kong, January 2008.
507. International Conference on Retrofitting of Structures, MCEER, Columbia University, New York, USA, March 1996.
508 Life Cycle Assessment of Sustainable Infrastructure Materials, US -Japan Workshop, Editors N. Grace, M. Lepech, T. Ueda, Hokkaido University, Sapporo, Japan, October 2010. http://www.hucc.hokudai.ac.jp/~m16120/workshop2009/US-Japan.Report.Final.pdf
509. Life Cycle Performance of Deteriorating structures – Assessment, Design & management, edited by D. M. Frangopol, E. Brühwiler, M. H. Faber, B. Adey , 3rd IABMAS Workshop on Life-Cycle Cost Analysis and Design of Civil Infrastructure Systems (LCC3) & JCSS Workshop on Probabilistic Modelling of Deterioration Processes in Concrete Structures, Lausanne, March 2003, ASCE Press, 2004.
510. Maintenance and Asset Management your key Recovery Drivers – The Theme for the 20th Euro maintenance International Conference and Exhibition held in Verona Italy May 12–14, 2010.
511. Operation, Maintenance and Rehabilitation of Large Infrastructure Projects, Bridges and Tunnels, IABSE Conference, Copenhagen, Denmark, May 2006.
512. Predicting the Performance of Concrete Repair Materials, by A. M. Vaysburd, N. J. Carino, B. Bissonette, NISTIR 6402, Durham, New Hampshire, USA, April 1999.
513. Repair of Concrete Structures: From Theory to Practice in a Marine Environment, Norwegian Road Research Laboratory, Oslo, Editor: A. Blankvoll, May 1997.
514. Service Life and Serviceability of Concrete Structures, European Symposium, ESCS, Helsinki, Finland, June 2006.
515. Structural Assessment: The Role of Large and Full-Scale Testing, Editors: G. S. T. Armer; J. L. Clarke; F. K. Garas; K. S. Virdi, Routledge, IstructE, UK, 1997.

516. Structural Faults & Repair: Extending the Life of Bridges, Concrete + Composites, Buildings, Masonry + Civil Structures, 12th International Congress and Exhibition, Edinburgh, Scotland, June 2008. www.structuralfaultsandrepair.com
517. Structural Faults and Repair, Proceedings of 7th International Conference, edited by M. C. Forde. Vol. 3, Extending the life of buildings and civil structures, Engineering Technics Press, Edinburgh, Scotland, 1997.
518. The Need to Relate Fabric Deformations to both the Soil Conditions and History of the Building, by A. Lord,) Symposium on Building Appraisal, Maintenance & Preservation, Department of Architecture and Building Engineering, University of Bath, UK, July 1985.

Note: References mentioned above present a partial list of specialized conferences related of the topics of SED 12. Several other excellent conferences were also organized which might not be mentioned in this partial list. Moreover, many of the conferences on Structural Engineering topics include special sessions on Assessment, Inspections, Evaluation, Conservation, Maintenance, Preservation, Repair, Rehabilitation, Retrofitting, Strengthening, Upgrading, . . . of Structures

(VI) References on Historical and Heritage Structures (Partial List):

519. AASHTO – Historic Preservation / Cultural Resources. http://environment.transportation.org/environmental_issues/historic_cultural/docs_reports.aspx# bookmarksubHistoricBridgeManagementandTreatmentPlans
520. Architecture of Additions: Design and Regulation, W.W. Norton & Co., New York, 1998.
521. ASTM C1496-09 – Standard Guide for Assessment and Maintenance of Exterior Dimension Stone Masonry Walls and Facades, American Society for Testing and Materials, USA, 2009.
522. ASTM STP1355 – The Use of and Need for Preservation Standards in Architectural Conservation, Edited by L. B. Sickels-Taves, American Society for Testing and Materials, 1999.
523. CARE – ICE – The Conservation Accreditation Register for Engineers (CARE) has been established to identify engineers skilled in the conservation of historical structures and sites, ICE, Institution of Civil Engineers, UK. http://www.ice.org.uk/care
524. CIRIA R111 – Structural Renovation of Traditional Buildings, Construction Industry Research and Information Association, 1986.
525. Cleaning Historic Buildings. Vol. 2. Cleaning Materials and Processes, by N. D. Ashurst, Shaftesbury, 1994.
526. Computational Modeling of Masonry, Brickwork and Blockwork Structures, by J. W. Bull, Saxe-Coburg Publications, UK, 1997.
527. Conservation of Historic Brick Structures, by N. S. Baer, S. Fitz and R.A. Livingston, UK, 1998
528. Conservation of Historic Buildings, by S. B. M. Feilden, Butterworth-Heinemann, London, 2003.
529. Conservation of Timber Buildings, by F. W. B. Charles, and M. Charles, Stanley Thornes Publisher Ltd., 1990.
530. Conserving Buildings: Guide to Techniques and Materials, Revised Ed. M. E. Weaver, John Wiley & Sons Inc, 1993.
531. Corrosion in Masonry-Clad Early 20th-Century Steel-Framed Buildings, by P. Gibbs, TAN 20. Historic Scotland, Edinburgh, 2000.
532. Dictionary of Building Preservation, Editors: W. Bucher, C. Madrid, John Willy & sons, Inc., Canada,1996
533. English Heritage Graffiti on Historic Buildings and Monuments. Methods of Removal and Prevention. Technical Advice Note. English Heritage, London, 1999.
534. FHWA – National Historic Covered Bridge Preservation Program http://www.fhwa.dot.gov/bridge/covered.cfm
535. Guidelines for Historic Bridge Rehabilitation & Replacement, AASHTO GHBRR-1, 1st Ed., USA, 2008.
536. Heron – Maintenance and repair of natural stone, Editors T.G. Nijland, R.P.J. van Hees, Vol. 54, No. 4, Delft, The Netherlands, 2009. http://heron.tudelft.nl/54-4/54-4.html

537. Historic Building Façades: The Manual for Maintenance and Rehabilitation, by W. G. Foulks (Editor), J. M. Fitch (Foreword), New York Landmarks Conservancy, Wiley and Sons, 1997.
538. Historical Constructions 2001, Possibilities of numerical and experimental techniques, Edited by P. B. Lourenço & P. Roca, of 3rd Int. Seminar, Guimarães, Portugal, 2001.
539. IABSE – Colloquium on History of Structures, 1982.
540. ICOMOS – Principles for the Analysis, Conservation and Structural Restoration of Architectural Heritage, International Council on Monuments and Sites, France, 2003. http://www.international.icomos.org/charters/structures_e.htm
541. ICOMOS – Recommendations for the Analysis, Conservation and Structural Restoration of Architectural Heritage, France, 2003. http://www.civil.uminho.pt/masonry/Publications/Recommendations_ICOMOS.pdf
542. ICOMOS – The Burra Charter, The Australia ICOMOS Charter for the Conservation of places of Cultural Significance, France, 1999.
543. ICOMOS – The Venice Charter, International Charter for the Conservation & Restoration of Monuments and Sites, France, 1964.
544. International Symposium on the Concepts and Practices of Conservation and Restoration of Historic Buildings in East Asia, Beijing, China, May, 2007. http://www.sach.gov.cn/tabid/230/InfoID/7022/Default.aspx
545. Masonry: How to Care for Old and Historic Brick and Stone, by M. London, National Park Service Staff (Editors), Preservation Press, 1988.
546. Modern Matters: Principles and Practice in Conserving Recent Architecture, Edited by S. Macdonald, Proceedings of the English Heritage Conference, London, UK, 1995.
547. NCCR – National Council for Conservation-Restoration.
http://cool.conservation-us.org/byform/mailing-lists/cdl/2002/1164.html
548. NPS – A Checklist For Rehabilitating Historic Buildings, National Park Service, USA, 2009 http://www.nps.gov/history/hps/tps/cheklist.htm
549. NPS – Guidelines for Local Surveys: A Basis for Preservation Planning, Appendix 4, Bibliography, National Park Services, USA, 1985. http://www.nps.gov/history/nr/publications/bulletins/nrb24/appendix4.htm
550. NPS – Historic Preservation: Action Transforming Our Thinking, National Park Services, USA, 2008. http://www.nps.gov/history/history/resedu/historic_preservation.pdf
551. NPS – National Park Services, Preservation, http://www.nps.gov/history/preservation.htm
552. NPS – Preservation of Historic Concrete: Problems and General Approaches, by W. Coney, Preservation Brief No. 15, US Department of the Interior, National Park Service, Washington, DC, 1996.
553. NPS – Standards for Rehabilitation, The Secretary of the Interior's Standards for the Treatment of Historic Properties, National Park Service, 1990.
554. NPS – The Secretary of the Interior's Standards for the Treatment of Historic Properties with Guidelines for Preserving, Rehabilitating, Restoring & Reconstructing Historic Buildings, K. D. Weeks and A. E. Grimmer, U.S. Department of the Interior, National Park Service, USA, 1995. http://www.nps.gov/history/hps/tps/standards_guidelines.htm
555. NPS-28 – Cultural Resource Management Guideline, 1998.
Appendix L: Historic and Prehistoric Structure Technical Bibliography, National Park Service, (a) Preservation Briefs, (c) Preservation Tech Notes, Preservation Case Studies, Videotapes. http://www.nps.gov/history/history/online_books/nps28/28appenl.htm
556. NSF – International Workshop on Preservation of Historical Structures With FRP Composites, by L. De Lorenzis & A. Nanni, National Science Foundation (NSF), USA, 2004. http://www.iifc-hq.org/publications/reports.html
557. NTHP – Building Codes and Historic Buildings, by M. Green, Anne Watson, Item 2B15, National Trust for Historic Preservation, Washington DC, USA, 2005.
558. NTHP – Case Studies, National Trust for Historic Preservation, http://www.preservationnation.org/resources/case-studies/
559. NTHP – Respectful Rehabilitation: Answers to Your Questions About Old Buildings, National Trust for Historic Preservation, Washington DC, 1982.
560. NTHP – The Economics of Historic Preservation: A Community Leaders' Guide, National Trust for Historic Preservation, Washington DC, 1994.

561. Practical Building Conservation, by J. Ashurst, and N. Ashurst, English Heritage Technical Handbook, Vol. 1: Stone Masonry, Vol. 2: Brick, Terracotta and Earth and Vol. 3: Mortars, Plasters and Renders, Gower Technical Press, 1988.
562. Preservation Comes of Age: From Williamsburg to the National Trust, 1926-1949, Charlottsville, University Press of Virginia, 1981.
563. Preservation, conservation & restoration of buildings and monuments, UK, 2004. http://www2.rgu.ac.uk/schools/mcrg/stpreserv.htm
564. Preserving the Recent Past, Vol. 1, Editors: D. Slaton, R. Shiffer, G. Foulks, G., Historic Preservation Education Foundation, USA, 1995.
565. Preserving the Recent Past, Vol. 2, Editors: D. Slaton, and G. Foulks, Historic Preservation Education Foundation, USA, 2000.
566. Professional Guidelines for Assuring and Preserving the Authenticity of Heritage Sites in the Context of the Cultures of Asia, HOI An Protocols for Best Conservation Practice In ASIA, by R. A Engelhardt, P. R. Rogers, Prepared for UNESCO, Adopted by the Asia-Oceania Region at the ICOMOS General Assembly in Xi'an, China, 2005. http://unesdoc.unesco.org/images/0018/001826/182617e.pdf
567. Retrofitting Suburbia: Urban Design Solutions for Redesigning Suburbs, by E. Dunham-Jones, J. Williamson, John Wiley & Sons, 2009.
568. RILEM – Proceedings pro062 – In-Situ Evaluation of Historic Wood and Masonry Structures, Edited by B. Kasal, R. Anthony and M. Drdacky, International Union of Laboratories and Experts in Construction Materials, System and Structures, 2009.
569. RILEM – Proceedings pro067 – Repairs Mortars for Historic Masonry, Edited by C. Groot, 2009.
570. RILEM Report rep 028 – Characterisation of Old Mortars with Respect to their Repair – Final Report of RILEM TC 167-COM, Edited by C. Groot, G. Ashall and J. Hughes, 2004.
571. Staining of Facades, by M. Y. L. Chew & T. P. Ping, World Scientific Pub Co Inc., 2003.
572. Standards for Preservation and Rehabilitation, by S. J. Kelley, ASTM, 1996.
573. Structural Analysis of Historic Buildings: Restoration, Preservation, and Adaptive Reuse Applications for Architects and Engineers, J. Stanley Rabun, Wiley, 2000.
574. Structural Aspects of Building Conservation, by P. Beckmann, R. Bowles, McGraw-Hill Companies 2nd Ed., 2004.
575. Structural Investigation of Historic Buildings: A Case Study Guide to Preservation Technology for Buildings, Bridges, Towers and Mills, by D. C. Fischetti, Wiley, 2009.
576. Structural Masonry, by A. W. Hendry, Macmillan Education Ltd., 1990.
577. Structural Studies, Repairs and Maintenance of Heritage Architecture XI, 11th Conference, Edited by C.A. Brebbia, WIT Press, UK, 2009.
578. Structure and Style: Conserving Twentieth-Century Buildings, by M. Stratton, (ed) E & FN Spon, London, UK, 1997.
579. Surveying Historic Buildings, by P. Swallow and D. Watt, Donhead Publishing, 1996.
580. The Building Conservation Directory. http://www.buildingconservation.com/
581. The Conservation and Structural Restoration of Architectural Heritage, by G. CROCI, WIT Press, UK, 1998.
582. The Engineering Geology of Ancient Works, Monuments, and Historical Sites: Preservation and Protection, Edited by P. G. Marinos and G. C. Koukis, A. A. Balkana, Rotterdam, Brookfield, Athens, 19-23 September 1988.
583. The Powers of Preservation : New Life for Urban Historic Places, A. C. Moore, McGraw-Hill; 1998.
584. The Protection of Historic Properties: A Comparative Study of Administrative Policies, by C. Sanz, WIT Press, UK, 2009.
585. The Secretary of the Interior's Standards for the Treatment of Historic Properties, by K. D. Weeks and Anne E. Grimmer, Washington, D.C., 1995.
586. The Society for the Protection of Ancient Buildings. http://www.spab.org.uk/
587. Urban Planning Conservation & Preservation, N. Cohen, Architectural Press, 1982.
588. Wood Preservation, by B. A. Richardson, Construction Press, 2nd Ed., E. & F.N.Spon, UK, 1993.

(VII) Videos and Presentations (Partial List):

589. Concrete Defects & Repair Strategy, by ECY Yiu, Department of Real Estate & Construction, Presentation, January 2007. ww.hkis.org.hk/hkis/html/upload/CPDLink/2007034.pdf
590. Concrete Network – Concrete Repair Video Series. http://www.concretenetwork.com/concrete-repair-videos/
591. FHWA – Pennsylvania Bridges: Maintaining the Past-Preserving the Future, Construction and Maintenance Fact Sheets, Managing Highway Assets: Bridge Preservation, Federal Highway Administration, Video. http://www.fhwa.dot.gov/bridge/preservation/pennwm.wmv
592. IABSE E-Learning – Refer to Appendix A in this document, Section No. IV, for a list of Lectures (Audio Visual presentations). http://www.iabse.org/
593. NCPP – FP2 Pavement Preservation Video, National Center for Pavement Preservation, USA. http://www.pavementpreservation.org/video/index.php
594. NPS-28 – Cultural Resource Management Guideline, Appendix L: Historic and Prehistoric Structure Technical Bibliography, National Park Service, Videotapes. http://www.nps.gov/history/history/online_books/nps28/28appenl.htm
595. RTA-NSW – Historic Timber Bridge Maintenance, Road and Traffic Authority (RTA) – NSW Government, Australia, Video. http://www.rta.nsw.gov.au/environment/heritage/rtaoralhistory program/timber_bridge_maintenance.html
596. WHE –World Housing Encyclopedia Website Library, Animations. http://world-housing.net/index.php

(VIII) Relevant Websites & Relevant References (Partial List):

Part (1): Position Papers, Policy Statements, Strategies

597. AASHTO BTG-1 – Bridging the Gap: Restoring and Rebuilding the Nation's Bridges, American Association of State Highway and Transportation Officials, USA, 2008. http://www.transportation1.org/bridgereport/docs/BridgingtheGap.pdf
598. AASHTO TSP2 – Transportation System Preservation Technical Services Program (TSP•2) http://www.tsp2.org/library
599. ASCE – Report card for America's infrastructure, American Society of Civil Engineers, 2007. http://www.asce.org/reportcard/2005/index2005.cfm
600. ASCE Report Card for America's Infrastructure, 2009. http://www.infrastructurereportcard.org/report-cards
601. FHWA – Asset Management Position Paper, Federal Highway Administration, Washington DC, USA, 2008. http://www.fhwa.dot.gov/infrastructure/asstmgmt/positionpaper.cfm
602. FHWA – The Bridge Preservation and Maintenance (BPAM) Roadmap. http://www.fhwa.dot.gov/bridge/preservation/roadmap2007.cfm
603. ICE – Coordination of Infrastructure Development: Some International Comparisons http://www.ice.org.uk/Information-resources/Document-Library/Co-ordination-of-infrastructure-development–Inter
604. JSCE – Maintenance of Concrete Structures and its Future Strategy – Based on Newest Concepts of the Maintenance Part in the JSCE Standard Specification for Concrete Structures, by K. Takewaka and Y. Kaneko, 2006. http://www.jsce.or.jp/committee/concrete/e/newsletter/newsletter07/KSCE_Takewaka.pdf
605. NCN – National Composites Network (NCN), UK. Repair of Fibre Reinforced Polymer (FRP) Structures – NCN Best Practice Guide.
http://www.ncn-uk.co.uk/DesktopDefault.aspx?tabindex=111&tabid=400
NCN Case Studies. http://www.ncn-uk.co.uk/DesktopDefault.aspx?tabindex=16&tabid=315
606. Rethinking Schools Capital Investment: The New 3Rs? Refresh, Refurbish, Reuse.
http://www.thecentreforschooldesign.org/wp-content/uploads/2010/06/TheNew3Rs-1.pdf
(British Council for School Environments, London, 2010) part 1

607. TCRP Digest 92 – Balancing Infrastructure Reinvestment with System Expansion, TCRP Research Results Digest, TRB's Transit Cooperative Research Program, 2009. http://onlinepubs.trb.org/onlinepubs/tcrp/tcrp_rrd_92.pdf

Part (2): Online Courses (Partial List)

608. ESDEP WG 16 – Structural Systems: Refurbishment, Lecture 16.1: Strengthening of Structures, European Steel Design Education Programme. http://www.esdep.org/4ccr/members/master/wg16/t0100.htm
609. PDH – Online Course Library, Structural Engineering.
http://pdhonline.org/cgi-bin/quiz/courses/listings.cgi?category_id=5
(A105) Preservation, (A 106) Rehabilitation, (A 107) Restoration, (A 108) Reconstruction Standards and Guidelines for Historic Buildings; (A110) The Preservation & Repair of Historic Stained and Leaded Glass; (A113) Repair and Replacement of Historic Slate Roofs; (A114) Preservation of Historic Adobe Buildings; (S136) Evaluation and Repair of Concrete Structures; (S155) Concrete Deterioration; (S173) Concrete Maintenance Guidelines; (S175) Concrete Repair – Selection of Repair Methods; (S176) Concrete Removal and Repair – Methods and Materials; (S177) Petrographic Analysis of Concrete Deterioration; (S211), (S 212) Antiquated Structural Systems – Part 1, Part 2.

Part (3): Technical societies, Associations, Research Samples, Resource Websites (Partial List), (please refer also to section III above)

610. ACMA – American Composites Manufacturers Association. http://www.mdacomposites.org/mda/Site_Map.htm
External Reinforcement Systems – Concrete Repair, Strengthening & Seismic Retrofit
http://www.mdacomposites.org/mda/psgbridge_concreterepair_intro.html
Codes and Standards http://www.mdacomposites.org/mda/codes_standards.html
611. Building Magazine http://www.buildings.com/Magazine/tabid/3070/Default.aspx
612. Composite Materials Links http://www.composites.ugent.be/links.html
613. Concrete Restoration and Concrete Repair Guide. http://www.onlineconcreterestoration.com/
614. EFNMS – European Federation of National Maintenance Societies. http://www.efnms.org/
615. FHWA – Asset Management. http://www.fhwa.dot.gov/infrastructure/asstmgmt/
616. FHWA – Bridge Management. http://www.fhwa.dot.gov/bridge/management/index.cfm
617. FHWA – Bridge Technology. http://www.fhwa.dot.gov/bridge/
618. FHWA – Preservation – Bridges. http://www.fhwa.dot.gov/bridge/preservation/
619. FHWA – Transportation System Preservation. http://www.fhwa.dot.gov/preservation/index.cfm
620. FHWA ASR – Alkali-Silica Reactivity, Development and Deployment Program http://www.fhwa.dot.gov/pavement/concrete/asr.cfm
621. FHWA NBI – National Bridge Inventory http://www.fhwa.dot.gov/bridge/nbi.htm
622. FHWA NBIS – National Bridge Inspection Standards http://www.fhwa.dot.gov/bridge/nbis.htm
623. FHWA TEA-21 – Transportation Equity Act for the 21st Century.
– Highway Bridge Replacement and Rehabilitation Program, 1998.
http://www.fhwa.dot.gov/tea21/factsheets/r-rrehab.htm
– Interstate Maintenance Program – Resurfacing, Restoring, Rehabilitating and Reconstructing (4R), 1998. http://www.fhwa.dot.gov/tea21/factsheets/im.htm
624. Good Management of Road Capital. http://www.mlit.go.jp/road/road_e/contents01/1-3-3.html
625. IABMAS – International Association for Bridge Maintenance & Safety. http://www.iabmas.org/
626. IHBC – Institute of Historic Building Conservation. http://www.ihbc.org.uk/
627. IIFC – International Institute for FRP in Construction. http://www.iifc-hq.org/
628. InspectAPedia – InspectAPedia – Building & Environmental Inspection, Testing, Diagnosis. Repair, http://www.inspectapedia.com/
– Foundation Repair Methods – Examples of Typical Foundation Repairs http://www.inspectapedia. com/structure/FoundationRepair.htm

– How to Evaluate Cracks in Poured Concrete Slabs & Floors. http://www.inspectapedia.com/structure/FloorCracks.htm
– Repair Methods and Products for Damaged Foundations, walls, & slabs
http://www.inspectapedia.com/structure/FoundationRepair.htm
– Structural Defects, Inspection, Diagnosis, & Repair. Rot & Insect Damage to Buildings. http://www.inspectapedia.com/structure/structure.htm

629. ISHMII – International Society for Structural Health Monitoring of Intelligent Infrastructure. http://www.ishmii.org/Literature/SHMGlossaryDefinitions.html
630. ISIS – Intelligent Sensing for Innovative Structures, Canada. http://www.isiscanada.com/
631. Journal of Construction Technology of China. http://www.shigongjishu.cn/
632. MQA – Maintenance Quality Assurance – Document Library. http://www.mrutc.org/outreach/MQA/library/
633. NBM Media – Construction information, Repairs & Rehabilitation, India. http://www.nbmcw.com/articles/repairs-a-rehabilisation.html
634. NCEF – School Modernization and Renovation, National Institute of Building Sciences. http://www.edfacilities.org/rl/renovation.cfm# sites
635. NCN – National Composites Network (NCN), UK. Repair of Fibre Reinforced Polymer (FRP) Structures – NCN Best Practice Guide.
http://www.ncn-uk.co.uk/DesktopDefault.aspx?tabindex=111&tabid=400
NCN Case Studies. http://www.ncn-uk.co.uk/DesktopDefault.aspx?tabindex=16&tabid=315
636. NISEE – National Information Service for Earthquake Engineering. http://nisee.berkeley.edu/
637. Restoration Online: http://www.restorationonline.com/
638. Retrofit Solutions Project Overview, New Zealand. http://www.retrofitsolutions.org.nz/index.shtml Seismic Retrofit Solutions NZ: Latest News. http://www.retrofitsolutions.org.nz/pdfs/Ingham%20June%2006.pdf
639. Structural Engineer, Repair and Retrofit of Structures. http://www.thestructuralengineer.info/library/repair_structures.html
640. SWRI – Sealant, Waterproofing and Restoration Institute. http://www.swrionline.org/
641. TFRC – Long-Term Bridge Performance Program http://www.tfhrc.gov/ltbp
642. The Structural Engineer, A Center for Integrating Information on Structural Engineering. Section on: Repair and Retrofit of Structures. http://www.thestructuralengineer.info/onlinelibrary/index.php?classification=7
643. Today's Concrete Technology, Concrete Repair. http://www.todaysconcretetechnology.com/category/concrete-repair
644. TRB – Publications about Maintenance and Preservation. http://www.trb.org/Maintenance Preservation/TRBPublications.aspx
645. WBCSD – World Business Council for Sustainable Development – Case Studies. http://www.wbcsd.ch/templates/TemplateWBCSD5/layout.asp?MenuID=1
646. WHE – World Housing Encyclopedia. http://world-housing.net/index.php
647. World Wide Composite Search Engine. http://www.wwcomposites.com/

Part (4): Materials & Products (Partial List)

648. BASF – Master Builders. http://www.basf-admixtures.com/EN/Pages/default.aspx
649. BBR – Holdings Ltds.
Restoration, renovation and redevelopment/conversion works to existing buildings.
– http://www.bbr.com.sg/restoration.html
– Specialised Engineering – Maintenance Repair & Retrofitting
– http://www.bbr.com.sg/repair-retrofit.html
650. ChemCo Systems http://www.chemcosystems.com/descri.html
651. DSI – DYWIDAG Systems International. http://www.dywidag-systems.com/
652. Edge Structural Composites, Inc. http://www.edgest.com
653. Fortress Stabilization Systems – foundation repair and concrete repair solutions.
Case Studies http://www.fortressstabilization.com/casestudies/energynet.php

654. Freyssinet Group – Repair and maintenance. http://www.freyssinet.com/appli/internet/w3fcom.nsf/ag_Creation_Page?OpenAgent&UNID=28D1D028ABD356FAC12573B10044A835&contexte=121&rubrique=activites&lang=en&font=small
655. FYFE Company – FRP Strengthening Industry http://www.fyfeco.com/applications/ http://www.fyfeco.com/resources/Literature.aspx
656. Gordon Composites, Inc. http://www.gordoncomposites.com/markets_infrastructure.htm
657. HILTI – Civil Engineering Bridge Construction Maintenance & Repair http://www.hilti.com/holcom/page/module/product/prca_catnavigation.jsf?lang=en&nodeId=-8169
658. OVM – Liuzhou Ovm Machinery Co., LTD., P.R. China, Strengthening & Repairing. http://www.ovm.cn/ovmweb-e/EECenter/EEList.aspx?eeTypeId=&infoType=24
659. Saint-Gobain Technical Fabrics. http://www.sgtf.com/Applications/ConcreteReinforcement
660. SIKA – repair & Protection. http://www.sikaconstruction.com/con/con-repair.htm
Structural Strengthening. http://www.sikaconstruction.com/con/con-repair/con-repair-ss.htm
661. SPS – Concrete Repair – Case Studies. http://www.spsrepair.com/
662. STABL WALL – The Best Choice for Foundation Repair. http://www.stablwall.com/
663. VSL – Repair/Strengthening. http://www.vsl.net/applicationscasestudies/bystructuremarket/repairstrengthening/tabid/115/default.aspx
664. WEBER – http://www.netweber.co.uk/construction-mortar-products.html
Problem solution: http://www.netweber.co.uk/construction-mortar-products/how-to-apply-our-products/problem-solutions.html
Case study : http://www.netweber.co.uk/construction-mortar-products/find-the-right-solution/case-studies.html
Product selection: http://www.netweber.co.uk/construction-mortar-products/find-the-right-solution/product-selector.html

(IX) Checklists (Partial List):

Checklists could be a convenient tool for quality assurance & quality control in design, construction, inspection, maintenance & repair of structures. Below follows a listing of 'Checklists' relevant to the topics of SED 12.

665. ACI 201.1 R-08 –Guide for Conducting a Visual Inspection of Concrete in Service, ACI, 2008, Ch. 3: Visual inspection report and checklist
666. ACI 362.2R-00 – Checklist for structural inspection of parking structures from Guide for Structural Maintenance of Parking Structures. (Appendix C, p.13 – 15, 2000).
667. AGC – Concrete Pre-Construction Checklist, Developed by Georgia Concrete & Products Association (GC&PA) with Georgia Branch (AGC), 2008.
668. Bridge Maintenance Checklist, From Bridge Inspection Program Documents, Nebraska Department of Roads, http://www.dor.state.ne.us/design/bridge/bipm/forms/dr27.pdf
669. Samples of Checklist and Record of Specific Tasks Performed, from Code of Practice for Site Supervision, by TCP, Hong Kong, Buildings Department. (Appendix VI, p.80 to p.84), 2005.

Refer also to the following. Note: The following references were mentioned previously in Appendix B, hence only an abbreviation is presented here. Number between parentheses indicates the reference details.

- ATC-40 – Table 2.2 (p. 2-9 & 2-10) & Table 6-1. (p.6-28). (Refer to # 248)
- FEMA 310 – (Ch. 3: Screening Phase, Sec. 3.7: Structural Checklists, p.3-15), (Refer to # 317)
- FEMA P-420 – (Sec. 5.7 Seismic Evaluation, Fig. 5-7, p.58, 2009), (Refer to # 324)
- Handbook on Repairs and Rehabilitation of RCC Buildings, Ch. 4, Fig. 4.1, p. IV-3. (Refer to # 151)
- HUD – Residential Rehabilitation Inspection Guideline, Appendix E: Inspection Record, p. E-5 to E-21, 2000. (Refer to # 358)

- NCHRP 514 – (Sec. 2.2.7: QAP Checklists for FRP Construction, p. III-8 to III-34) (Refer to # 398)
- NCHRP 609 – (Sec. 2.2.7 QA Checklists for FRP Construction, p. B-9 to B-33), (Refer to # 401)
- NCPP – Pavement Preservation Checklist Series. http://www.fhwa.dot.gov/pavement/preservation/ppcl00.cfm (Refer to # 410)
- Periodic Structural Inspection of Existing Buildings, (Annex A: Checklist for Periodic Structural Inspection of Existing Buildings www.bca.gov.sg/PeriodicStructuralInspection/others/PSI_PE.pdf (Refer to # 82)

(X) Terminology (Partial List):

Several books, codes, reports, and websites include "definitions" and "terminology" relevant to the topics of SED 12. The following presents a listing of these references (partial list).

670. AASHTO – Bridge Terms Definitions. http://www.iowadot.gov/subcommittee/bridgeterms.aspx
671. AASHTO ATG-4-UL – AASHTO Transportation Glossary, 4th Ed., USA, 2009
672. ACI – Concrete Terminology, The American Concrete Institute, 2009. http://www.concrete.org/Technical/CCT/FlashHelp/ACI_Terminology.htm
673. ASCE TCLEE Monographs 22 – Seismic Screening Checklists for Water & Wastewater Facilities, by Editor: W. F. Heubach, 2003.
674. Concrete Terms. http://www.allmetalssupply.com/concrete_terms.htm
675. EN 1990: 2002 – Basis of structural design, Eurocode, UK 2002. (Sec. 1.5: Terms and definitions, P. 12 to 21)
676. Federal Standard 1037C. http://www.its.bldrdoc.gov/fs-1037/fs-1037c.htm
677. FHWA – Action Pavement Preservation Definitions (Memo). http://www.fhwa.dot.gov/pavement/preservation/091205.cfm
678. FHWA PD-96-001 – Recording and coding guide for the structure inventory and appraisal of the nation's bridge, Federal Highway Administration, Washington, D.C., USA, 1996.
679. ICC IEBC 2003 – International Existing Building Code, International Code Council, USA, 2003. (Ch. 2: Definitions)
680. ICRI – Concrete Repair Terminology 2010. www.icri.org/general/repairterminology2010.pdf
681. ISO 2394 – General principles on reliability for structures, Terms and definitions, 1998.
682. ISO 3898:1997 –Bases for design of structures – Notations – General symbols
683. ISO 6707-1:2004 – Building and civil engineering – Vocabulary – Part 1: General terms
684. ISO 6707-2:1993 – Building and civil engineering – Vocabulary – Part 2: Contract terms
685. Maintenance and Preservation A Definition, by A. O. King, P.E., CRAB (Country Road Administration Board) 2006. http://www.crab.wa.gov/LibraryData/Research_and_Reference_Material/Road_Maintenance/060206MaintenanceVSPreservation.doc.
686. Maintenance and Preservation – A Definition, by A.O. King, P.E., County Road Administration Board, 2006. http://www.crab.wa.gov/librarydata/research_and_reference_material/road_maintenance/060206maintenancevspreservation.doc.
687. Maintenance, repair, and operations (From Wikipedia). http://en.wikipedia.org/wiki/Maintenance,_repair,_and_operations
688. NPS DSC – Website Online Education, Workflows Website, National Park Service, Denver Service Center. (DSC Workflows Definitions), http://workflow.den.nps.gov/9_Glossary/glossary_r.htm
689. NPS National Park Service, Technical Preservation Service, The Secretary of the Interior's Standards for the Treatment of Historic Properties. http://www.nps.gov/hps/tps/standguide/
690. SIA 260 – Basis of Structural Design, by Swiss Society of Engineers and Architects, 2003. (Ch.1: Terminology)
691. The Burra Charter (The Australia ICOMOS charter for places of cultural significance), Australia ICOMOS, App. A, Articles: Definitions, 1999. http://www.nationaltrust.com.au/burracharter.html

Refer also to the following. Note: The following references were mentioned previously in Appendix B, hence only an abbreviation is presented here. Number between parentheses indicates the reference details.

- AASHTO MM-4 – Maintenance Manual for Bridges. (Sec. 3.1: Bridge Maintenance (Refer to # 8).
- API Std 653 – Tank Inspection, Repair, Alteration, & Reconstruction. (Sec. 3: Definition), (Refer to # 238).
- ASCE / SEI 41-06 – Seismic Rehabilitation of Existing Buildings. (P. 368: Definition) (Refer to # 21).
- BD 89/03 The Conservation of Highway Structures. (Ch. 1: Introduction: Definition) (Refer to # 34).
- Bridge and Highway Structure Rehabilitation and Repair, M. A. Khan, (p. xxii & xxiii), (Refer to # 88).
- Bridge Engineering Handbook, (Ch. 50: Introduction, F. W. Klaiber, T. J. Wipf) (Refer to # 208).
- Bridge Rehabilitation, W. Radomski, (Ch. 1: Introduction, Terminology & General Scope) (Refer to # 93).
- Computer Aided Rehabilitation of Sewer & Storm Water Networks, Saegrov. (Sec. 4.1.1), (Refer to # 105).
- Concrete Bridges: Inspection, Repair, Strengthening, Testing and Load Capacity Evaluation, V. K. Raina (Sec. 1.1: Introduction) (Refer to # 108).
- E-CFR – Title 23: Highways. (Part 650: Subpart C & D: Definitions) (Refer to # 35).
- EN 1504 – EN 1504 Part 1: Definitions & Part 9: (Sec. 3: Terms & definitions). (Refer to # 37).
- FHWA TEA-21 – Resurfacing, Restoring, Rehabilitating & Reconstructing (4R), 1998. (Refer to # 623).
- HUD-NAHB – Innovative Rehabilitation Provisions (P. A-2, Table A-1: Overview of NARRP for Repair, Renovation, Alteration, and Reconstruction) (Refer to # 361).
- HUD-NARRP – Nationally Applicable Recommended Rehabilitation Provisions, (Ch. 2: Definitions), (Refer to # 366).
- ICOMOS – Recommendations for the Analysis, Conservation and Structural Restoration of Architectural Heritage (Part III: Glossary) (Refer to # 541).
- ISO 13822 – Bases for Design of Structures (Sec. 3: Terms & Definitions) (Refer to # 61).
- JSCE Guidelines for Concrete No. 4 – Standard Specifications For Concrete Structures (Refer to # 67).
- Manual of Maintenance of Steel Bridge Structures: Planning, Design, and construction for Maintenance and Durability. (Sec. 1.2: Definitions of Terms) (Refer to # 73).
- NCHRP Web Document 35 – Rehabilitation Strategies for Highway Pavements. (Ch. 1: Definitions), (Refer to # 406).
- Professional Guidelines for Assuring and Preserving the Authenticity of Heritage Sites in the Context of the Cultures of Asia (Refer to # 566).
- Repair of Concrete Bridges, G. P. Mallett (Sec. Glossary) (Refer to # 174).
- Repair of Concrete Structures to EN 1504, Dansk Standard. (Sec. 3 Definitions and explanation of terms) (Refer to # 175).
- Seismic retrofitting of steel and composite building structures, Sarno & Elnashai. (Sec. 1.5. Definitions) (Refer to # 452).
- SIA 269 – Basics of Planning Structural Design, Swiss Code. (Ch. 1: Terminology) (Refer to # 80).
- Structural Renovation in Concrete, Z. Li, C. Leung, Y. Xi, (Sec. 1.4: Useful definition) (Refer to # 192).
- Structural Renovation of Buildings: Methods, Details, & Design Examples, A. Newman. (Sec. 1.1: Terminology) (Refer to # 193).

Appendix

C

Examples of Diagnostics of Crack Patterns and Causes of Deterioration in Concrete Structures

This Appendix includes examples of diagnostics crack patterns, deterioration patterns, and causes of deterioration in concrete structures.

Appendix C-1: Common causes of defects and general principles in EN 1504-9.
Appendix C-2: Diagnostic chart on symptoms and causes of cracking.
Appendix C-3a: A typical crack pattern of a portal frame pier with overhangs.
Appendix C-3b: A typical crack pattern of an abutment.
Appendix C-4: Schematic figure for material deterioration and performance degradation.
Appendix C-5a: Examples of cracking in reinforced concrete bridge deck slabs.
Appendix C-5b: Examples of cracking in reinforced concrete bridge beams.
Appendix C-5c: Examples of cracking in box girder bridges.
Appendix C-6: Common types of cracks and approximate time of appearance.
Appendix C-7: Examples of cracking in box girder bridges.

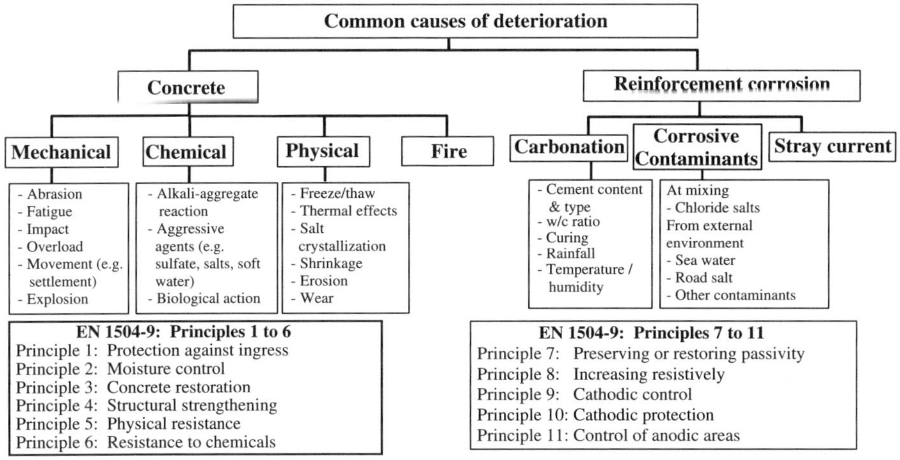

Refer also to Appendix D in this document

Appendix C-1: Common Causes of Defects and General Principles in EN 1504-9.
(Based on: EN 1504-9: Products and systems for the protection and repair of concrete structures. Principles for the use of products and systems, Figure 1)

General Cause		Specific Cause	Symptoms In:									
			Concrete						Steel		Member	
											Excessive Reflection	
			Active Cracks	Dormany Cracks	Spalling	Swelling	Discoloration	Surface Corrosion	Rusting of Steel	Yielding of Steel	Short Term	Permanent
1 Concrete Technology	1-1	Contaminated water (salt water oils)					X	X	X			
	1-2	Contaminates in concrete (sawdust)					X	X	X			
	1-3	Inappropriate admixtures (low F'c)			X	X						
	1-4	Expansive aggregates	X		X	X						
	1-5	Alkali-aggregate reaction						X	X			
	1-6	Poor surface finishing					X	X	X			
	1-7	Permeable concrete			X			X				
	1-8	Low temperature during pour						X				
2 Environmental Effects	2-1	Carbonic acid (dairies, …)						X				
	2-2	Hydrogen Sulfide (sewers)						X				
	2-3	Sulphate attack				X		X				
	2-4	Flue cases					X	X				
	2-5	Wave action water flow cavitations						X				
	2-6	Cycles of freezing and thawing						X				
	2-7	Cyclic wetting and drying				X		X				
	2-8	Cyclic chance in temperature (diurnal and / or seasonal)	X	X								
	2-9	Electrolytic attack	X		X							
3 Foundation Problems	3-1	Settlement										X
	3-2	Differential settlement	X	X	X							
	3-3	Lateral ground movement	X	X	X							
	3-4	Heaving	X	X	X							
	3-5	Frost			X			X	X			
4 Construction Errors	4-1	Formwork problems (lack of strength and stiffness. Lack of bracing premature removal)		X	X							X
	4-2	Poor concrete strength (added water batching errors incorrect or inappropriate mis. Etc)	X		X			X			X	X
	4-3	Inadequate curing of concrete (low strength high shrinkage high creep)	X		X							X
	4-4	Excessive		X				X				
	4-5	Oversize pours		X	X							
	4-6	Incorrect grade of steel	X							X		X
	4-7	Incorrect location and Quantity of Steel	X							X	X	X
	4-8	Overloads curing construction		X	X					X		X
5 Design Errors	5-1	Load under estimated	X		X					X	X	X
	5-2	Strengths over-estimated	X		X					X		X
	5-3	Stiffness over-estimated	X								X	X
	5-4	Omission of control joints	X	X								
	5-5	Inaccurate treatment of inelastic effects	X	X								X
	5-6	Inaccurate or inappropriate method of structural Analysis	X		X					X	X	X
6 Detailing Errors	6-1	Errors in transfer of information to drawings	X		X					X	X	X
	6-2	Steel congestion		X								
	6-3	Sharp chances in dimension along members	X	X	X							
	6-4	Bar termination	X	X								
	6-5	Inadequate cover		X					X			
	6-6	Anchorage problems	X	X	X							
7 Building Technology	7-1	Temperature effects (cold storage, air-conditioning)	X									
	7-2	Cladding (method of connection to frame, strength and durability)		X								
	7-3	Inadequate drainage (emission of drainage gradients: inadequate drains, drop groove omitted)					X	X				
	7-4	Inadequate wearing surface (for fork –lift trucks, etc)						X				
8 Unexpected Loads	8-1	Earthquake	X	X	X					X		X
	8-2	Gas explosion, some blast, explosive dust	X	X	X					X		X
	8-3	Collision	X	X	X					X		X
	8-4	Collapse of adjacent buildings	X	X	X					X		X
	8-5	Shock loads (standing waves, tension forces in concrete)	X	X	X					X		X

Appendix C-2: Diagnostic chart on symptoms and causes of cracking
(From: Strengthening, Stiffening and Repair of Concrete Structures, by R. F. Warner, IABSE SURVEYS S-17/81, IABSE Periodica 2/1981, page 42 – Table 2)

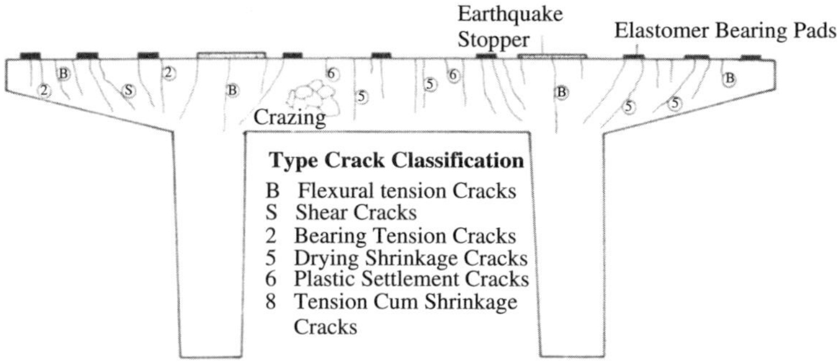

Appendix C-3a: A typical crack pattern of a portal frame pier with overhangs (or columns with a cross beam)

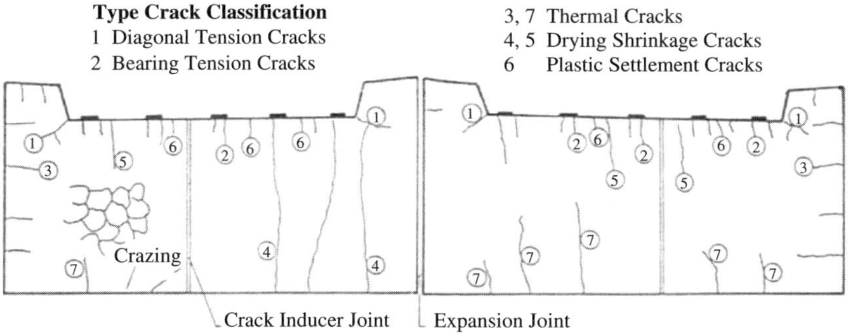

Appendix C-3b: A typical crack pattern of an abutment. (From: Deterioration of Reinforced Concrete Structures in a Marine Environment by D. V. Mallick, M. M. Tawil, IABSE, SEI 3, 1992)

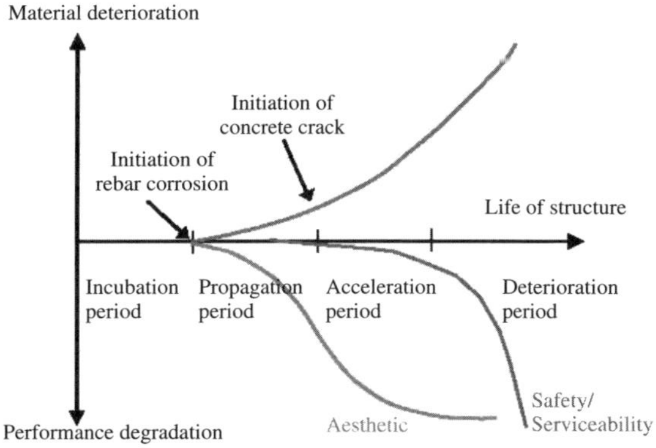

Appendix C-4: Schematic figure for material deterioration and performance degradation (From: Concept of Maintenance Part in the JSCE Standard Specifications for Concrete Structures and Its Future Strategy, by Koji Takewaka&ToyoakiMiyagawa, IABSE – fib Conference: Codes in Structural Engineering, Developments and Needs for International Practice, Croatia, May 3–5, 2010, Figure 6) & (From: Performance-based Standard Specifications for Maintenance and Repair of Concrete Structures in Japan, by Tamon Ueda, Koji Takewaka, IABSE SEI, 2007, Vol. 4, pp 359–366, Table 2)

Cracks in Skew Slabs (page 28)

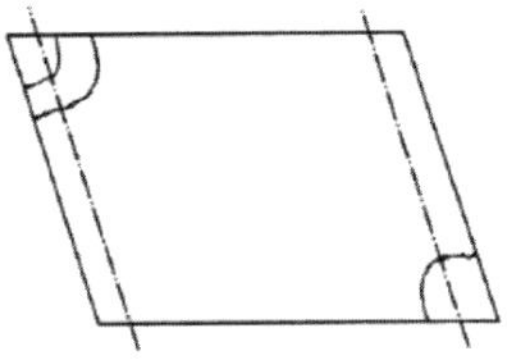

Cracks near Bearing edges (page 30)

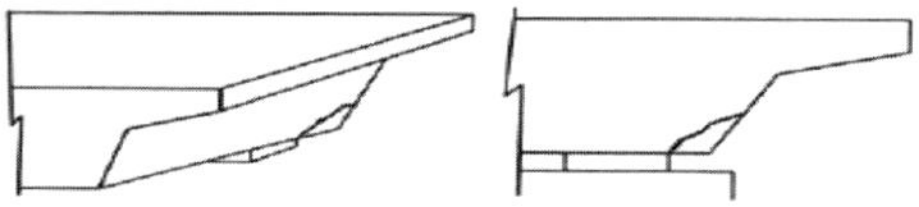

Appendix C-5a: Examples of cracking in reinforced concrete bridge deck slabs

Page (4) Vertical Cracks in Beams

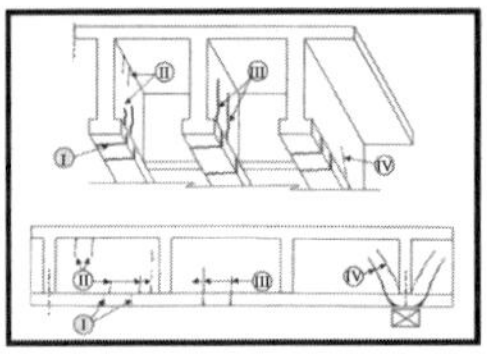

Page (26): Cracks in Cantilever Slabs

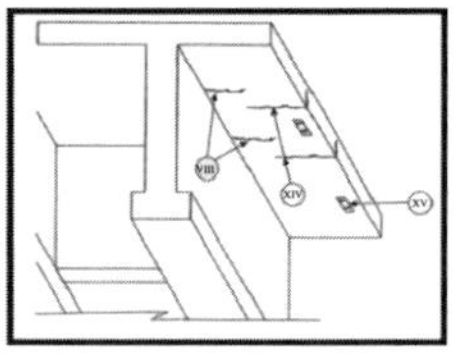

(page 14) Horizontal Cracks in Beams

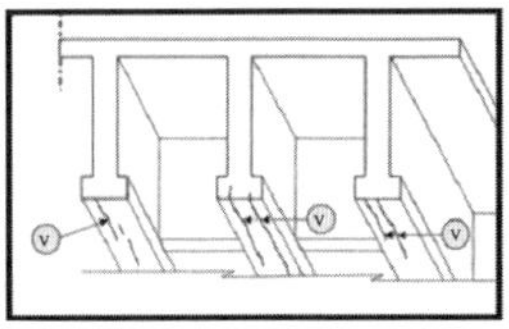

Page (20) Cracks in Beams & Concrete Spalling

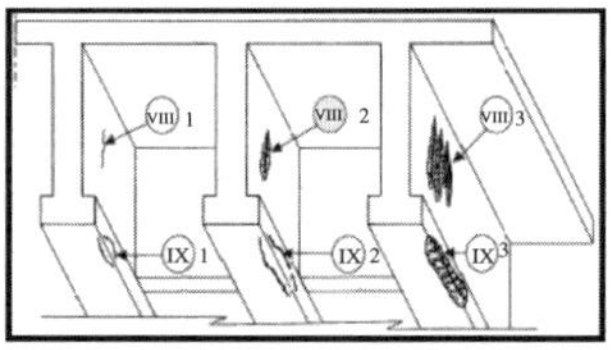

Appendix C-5b: Examples of cracking in reinforced concrete bridge beams

Flexure Cracks in Webs (page 14)

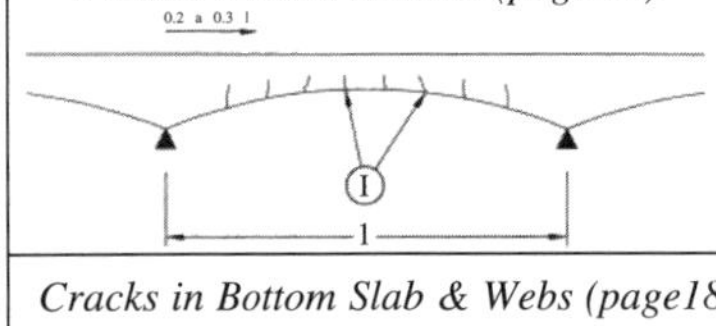

Shear Cracks in Webs (page 16)

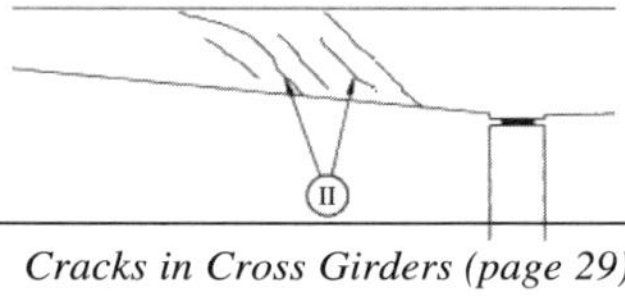

Cracks in Bottom Slab & Webs (page18)

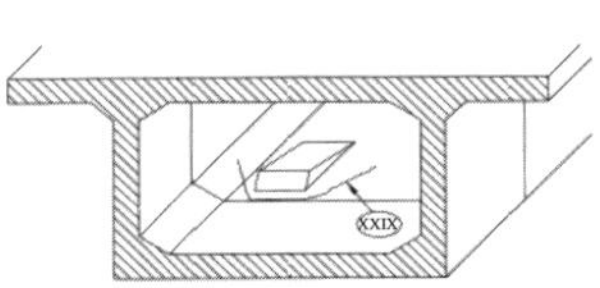

Cracks in Cross Girders (page 29)

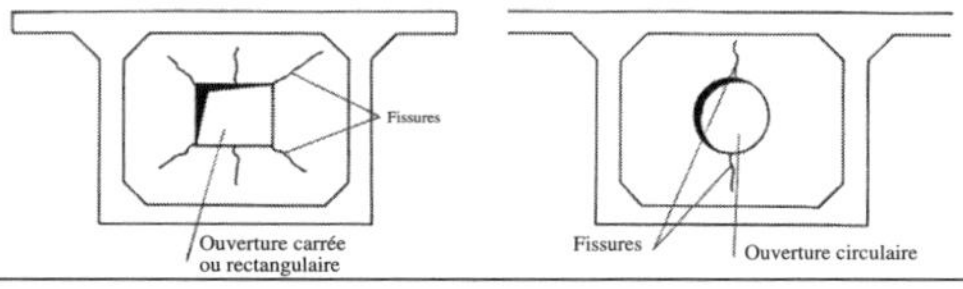

Appendix C-5c: Examples of cracking in box girder bridges
(From: SETRA, IQOA (Image Qualité des Ouvrages d'Art), Illustration des défauts: Poutre caisson en BP) Note: The Figures presented above provide a sample/example. For full details please refer to the website: http://www.piles.setra.equipement.gouv.fr/rubrique.php3?id_rubrique=33

Before Hardening	App. Time
Plastic Settlement: Rebar Cover, diameter, Mix, Slump	10 min–3 h
Plastic Shrinkage: Drying winds, Evaporation rate, Curing regime, Ambient temperatures, location of casting & construction joints	30 min–6 h
Early Frost Damage	
Construction: Formwork problems, Sub-Grade Movement	
Early Stages: Just after Hardening and just after Removal of Formwork	**App. Time**
Early thermal behavior & contraction: Effects of Internal restraint, External restraint, Heat of hydration, Dissipation rate, Differential effects	1 day–3 weeks
Surface Defects: Blowholes, Crazing, Honeycombing,	1 day–14 days
Construction: Early formwork or prop removal, Overloading	
After Hardening (Service Phase)	**App. Time**
Chemical	
External Sulfate Attack, Internal Sulfate Attack	1–5 years
Corrosion of Embedded Steel	3–10 years
Alkali–Silica Reactivity (ASR)	5–15 years
Alkali–Carbonate Reactivity (ACR)	5–15 years
Physical	
Drying Shrinkage: Evaporation of pore water, Relative humidity, Surface area, Water content, Curing	Weeks to months
Deicer Scaling and Deterioration	1–5 years
Freezing and Thawing Deterioration of Hardened Cement Paste, Aggregates	(1–5 years), (10–15 years)
Structural & Environmental	
Loads: Accidental Loads, Dynamic Loads, Overloading above design.	
Materials: Creep, Long-term shrinkage, Relaxation of post-tensioning steel, Fatigue	
Soil: Differential Settlement, soil heave, adjacent excavations	
Restraint of Movement: Long Term Drying shrinkage, Differential material effects, Location of joints, Thermal effects: Internal Temperature Gradients, External Seasonal Temperature Variations,	
Stress Concentrations	

Appendix C-6: Common types of cracks and approximate time of appearance

This Table is based on: several references, including:

- Non-Structural Cracks in Concrete, Concrete Society , Third Edition, 1992, Technical Report No. 22
- Control of Cracking in Concrete: State of the Art, Transportation Research Circular E-C107, 2006 http://www.trb.org/Construction/Blurbs/Control_of_Cracking_in_Concrete_State_of_the_Art_158019.aspx,
- Avoiding Surface Imperfections in Concrete: blowholes, crazing, dusting, flaking, honeycombing and popoutshttp://www.ccaa.com.au/publications/publication_search.php?searchtype=id&id=164
- Fundamentals of Durable Reinforced Concrete, M. G. Richardson, Taylor & Francis Group, 2002.

  Strengthening, Stiffening and Repair of Concrete Structures, by R. F. Warner, IABSE Surveys S-17/81, IABSE Periodica 2/1981.

Figure 7.11.27 Box girder cracks induced by shear

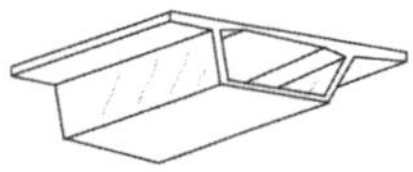

Figure 7.11.28 Box girder cracks induced by direct tension

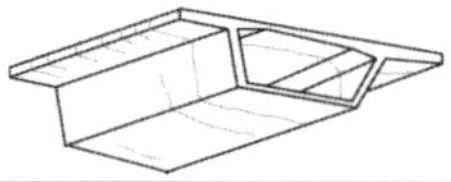

Figure 7.11.29 Box girder cracks induced by flexure (positive moment)

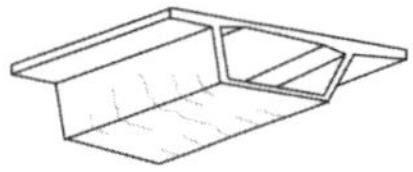

Figure 7.11.30 Box girder cracks induced by flexure (negative moment)

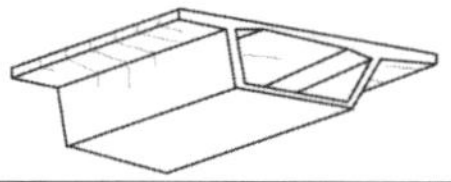

Appendix C-7: Examples of cracking in box girder bridges
(From: Bridge Inspector's Reference Manual, by T. W. Ryan, P.E, R. A. Hartle, J. E. Mann, L. J. Danovich, . Federal Highway Administration, BIRM, Volume 1, FHWA NHI 03-001, October 2002/November 2006 (page 7.11.21 & 7.11.22). Section7a- Inspection and Evaluation of Common Concrete Superstructures.)Note: The Figures presented above provide a sample/example. For full details please refer to the website: http://www.dot.state.mn.us/bridge/manuals/inspection/BIRM/BIRM_1.pdf

List of Some References on Cracking and Crack Patterns in Structures

1. Avoidance of thermal cracking in concrete at early ages, RILEM - Recommendations of TC 119-TCE, 1999.
2. Avoiding Early Cracking in Concrete, CCAA_ Concrete & Aggregates Australia. http://www.concrete.net.au/publications/publication_search.php?searchtype=id&id=53
3. Avoiding Surface Imperfections in Concrete: blowholes, crazing, dusting, flaking, honeycombing and popouts. http://www.ccaa.com.au/publications/publication_search.php?searchtype=id&id=164
4. Causes, Evaluation, and Repair of Cracks in Concrete Structures, ACI 224R-07, ACI Manual of Concrete Practice, American Concrete Institute, Farmington Hills, 2007.
5. Concrete Bridges: Inspection, Repair, Strengthening, Testing and Load Capacity Evaluation, by V. K. Raina, McGraw-Hill, 1994. Ch. 4: Cracks in Concrete—Types, Courses and Repair.
6. Concrete Repair and Maintenance Illustrated: Problem Analysis, Repair Strategy, Techniques, by P. H. Emmons, R. S. Means Company, USA, 2001.
7. Concrete Slab Surface Defects: Causes, Prevention, Repair, IS177, PCA, Portland Cement Association, Skokie, IL, USA, 2001.
8. Concrete Structures Stresses and Deformation, by A. Ghali; R. Favre; M. Elbadry, Spon Press, 3rd Edition, 2002. Chapter 7, 8, and 11: Stress, Strain, Displacements & Control of cracking.
9. Control of Cracking in Concrete Structures, ACI 224R-01, ACI, 2001.
10. Control of Cracking in Concrete: State of the Art, Transportation Research Circular E-C107, TRB, USA, 2006.
11. Crack Analysis in Structural Concrete: Theory and Applications, by Z. Shi, Butterworth-Heinemann, UK, 2009.
12. Crack and Crack Control at Concrete Structures, F. Leonhardt, IABSE Proceedings P-109/87, IABSE Periodica 1/1987.
13. Crack Control in Concrete Structures, F. Leonhardt, IABSE Surveys, No. S-4/77, IABSE, Zurich, 1977, page 26.
14. Cracking and building movement, by P. Dickinson, N. Thornton, RICS Books, 2004.
15. Cracking and Deformations, CEB Manual, EPFL, Lausanne, 1985.
16. Cracking Matters Journal, Concrete Repair Association, CRA, UK. http://www.concreterepair.org.uk/cracmatters.html
17. Defects and deterioration in buildings, by B. A. Richardson, Taylor & Francis, 2001.
18. Deterioration of Concrete: Symptoms, Causes, and Investigation (NRCC 44494), 2000. https://www.nrc-cnrc.gc.ca/eng/ibp/irc/catalogue/concrete-deterioration.html

19. Deterioration of Reinforced Concrete Structures in a Marine Environment by D. V. Mallick, M. M. Tawil, IABSE, SEI 3/92, page 193.
20. Diagnosis and Control of Alkali-Aggregate Reactions in Concrete, IS413T, by: J. Farny, S. Kosmatka, PCA, American Concrete Pavement Association, Skokie, IL, 1997.
21. Diagnostic Evaluation and Repair of Deteriorated Concrete Bridges, by A. Al-Ostaz, Conducted by The Department of Civil Engineering University of Mississippi in Cooperation with The Mississippi Department of Transportation, and U.S. Department of Transportation, FHA, December 2004.
22. Early Age Cracking in Cementitious Systems—Report of RILEM 025, Technical Committee 181-EAS, Early age shrinkage induced stresses and cracking in cementitious systems, Edited by A. Bentur, 2003.
23. Early Thermal Cracking of Concrete, BD 28/87, Design Manual for Roads and Bridges (DMRB). The Stationery Office, London, UK, Highways Agency. http://www.standardsforhighways.co.uk/dmrb/vol1/section3/bd2887.pdf
24. Early-age thermal crack control in concrete, by P. Bamforth, CIRIA C660, UK, 2007.
25. EN 1504: Products and systems for the protection and repair of concrete structures, Definitions, Requirements, Quality control and evaluation of conformity. Part 9: General principles for use of products and systems.
26. Fabrication and shipment Cracks in Prestressed Hollow-Core Slabs and Double Tees, PCI Journal Prestressed Concrete Institute, V. 28, No. 1, January–February 1983.
27. FHWA Bridge Inspector's Reference Manual, by W. Ryan, A. Hartle, J. Eric Mann, L. J. Danovich, Federal Highway Administration, *BIRM,* Volume 1, FHWA NHI 03-001, Oct. 2002/Nov. 2006, *Section7a- Inspection &Evaluation of Common Concrete Superstructures*(Fig. 7.11.27 to 7.11.37). http://www.dot.state.oh.us/Divisions/HighwayOps/Structures/bridge%20operations%20and%20maintenance/Pages/BridgeInspector%27sReferenceManual.aspx
28. FHWA Distress Identification Manual for the Long-Term Pavement Performance Program, John S. Miller and William Y. Bellingerm, (4th Ed.), FHWA-RD-03-031, USA, June 2003.
29. Fundamentals of Durable Reinforced Concrete, by M. G. Richardson, Taylor & Francis Group, 2002, pages 212–215.
30. Fundamentals of Earthquake Engineering, by A. S. Elnashaiandn L. Di Sarno, A John Wiley & Sons, 2008. Sec.: 1.4.1 Damage to Buildings and Lifelines & App. B: Damage to Structures.
31. Guide for Conducting a Visual Inspection of Concrete in Service, 201-1R-08, American Concrete Institute, 2008.
32. Guide to testing and monitoring the durability of concrete structures, CBDG Guide # 2, UK, 2002.
33. Movement, restraint and cracking in concrete structures, Technical Report No. 67, Concrete Society, UK, 2008.
34. Non-Structural Cracks in Concrete, Third Edition, Technical Report No. 22, Concrete Society, UK, 1992, pages 9 & 10.
35. Plastic and thermal cracking, by R. Day and J. Clarke, Chapter 2 in: Advanced Concrete Technology: Concrete Properties, Editors: John Newman, Ban SengChoo, Elsevier, 2003.
36. Presentation: Causes and Cures of Cracking in Concrete, by D. W. Fowler, J. J. King, Foundation Performance Association Houston, 2008. http://foundationperformance.org/pastpresentations/FowlerPresSlides-12Nov08.pdf
37. Presentation: Forensic Engineering: Causes of Distress in Concrete, by D. W. Fowler, Foundation Performance Association Houston, November 2009. http://www.foundationperformance.org/pastpresentations/FowlerPresSlides-11Nov09.pdf
38. Prestressed Concrete Structures, by M. P. Collins & D. Mitchell, Response Publications, Canada, 1997, pages 670-689.
39. Reinforced Concrete: Mechanics and Design, by J. G. MacGregor, Prentice Hall; 3rd Ed., 1996.
40. SETRA: IQOA (Image Qualité des Ouvragesd'Art), Presentations and Information on Cracking Patterns and Deterioration in Bridges, used for Visual Inspection of Bridges.

http://www.piles.setra.equipement.gouv.fr/rubrique.php3?id_rubrique=187http://www.piles.setra.equipement.gouv.fr/article.php3?id_article=367 SETRA: DCE, Presentations and Information on Repair of Bridges
http://www.piles.setra.equipement.gouv.fr/rubrique.php3?id_rubrique=178
- Slab bridges:
http://www.piles.setra.equipement.gouv.fr/IMG/pdf/Desordres_pont_dalle_ BA_cle051fdf.pdf
Also, http://www.piles.setra.equipement.gouv.fr/IMG/pdf/F9619PV_cle2131f8.pdf
- Reinforced Concrete Girder Bridges:
http://www.piles.setra.equipement.gouv.fr/IMG/pdf/Desordres_pont_a_poutres_en_BA_cle745b7c.pdf
Also, http://www.piles.setra.equipement.gouv.fr/IMG/pdf/F9621PV_cle25713c-2.pdf
- Prestresssed Box Girders :
http://www.piles.setra.equipement.gouv.fr/IMG/pdf/Desordres_pont_caisson_en_BP_cle5519ae.pdf

41. Specifications for Crack Repair–Position Statement No. 5, American Society of Concrete Contractors, ASCC, 2003.
42. Steel-Reinforced Concrete Structures: Assessment & Repair of Corrosion, by M. El-Reedy, CRC Press, 2007, pages 40, 41.
43. Strengthening, Stiffening and Repair of Concrete Structures, by R. F. Warner, IABSE Periodica, 2/1981, IABSE Surveys S-17/81.
44. The Cause of Cracking in Post-Tensioned Concrete Box Girder Bridges and Retrofit Procedures, by W.J. Podolny, Journal of Prestressed Concrete Institute, Vol. 30, No. 2, 1985, pages 82–139.
45. Types and Causes of Concrete Deterioration, PCA R&D No. 2617, USA, 2002.
46. Website: CCAA_ Concrete & Aggregates Australia. http://www.concrete.net.au/publications/
Documents available on the website include: Avoiding Early Cracking, Avoiding Surface Imperfections in Concrete, Blowholes (bug Holes), Chloride Resistance of Concrete, Compaction of Concrete, Concrete Basics, Concrete Pavement Maintenance/Repair, Curing of Concrete, Curling of Concrete Slabs, Delamination of Concrete Industrial Floors, Drying Shrinkage of Cement and Concrete, Dusting Concrete Surfaces, Early-age Shrinkage of Concrete, Efflorescence, Honeycombing, Joints in Concrete Buildings, Plastic Settlement Cracking, Plastic Shrinkage Cracking, Popouts, Slab Edge Dampness and Moisture Ingress, Sulfate-Resisting Cement and Concrete,. . ..
47. Website: NRMCA_ National Ready Mixed Concrete Association.
http://www.nrmca.org/aboutconcrete/cips/CIPseries.asp# trouble
Documents available on the website include: CIP1 Dusting Concrete Surfaces; CIP2 Scaling Concrete Surfaces; CIP 3 Crazing Concrete Surfaces; CIP 4 Cracking Concrete Surfaces; CIP 5 Plastic Shrinkage Cracking; CIP 7 Cracks in Concrete Basement; CIP8 Discrepancies in Yield; CIP9 Low Concrete Cylinder Strength Walls; CIP13 Concrete Blisters; CIP19 Curling of Concrete Slabs; CIP20 Delamination of Troweled Surfaces; CIP23 Discoloration; CIP25 Corrosion of Steel in Concrete; CIP40 Aggregate Popouts.
48. Website: PCA_ Troubleshooting / Repair, Frequently Asked Questions, Portland Cement. Association, Skokie, IL, USA http://www.cement.org/tech/cct_faqs.asp
Documents available on the website include: What causes concrete slabs to curl/warp, and how can this mechanism be minimized?, What causes random concrete cracks and can they be avoided?
49. Website: Today's Concrete Technology, Concrete Crack Repair Methods http://www.todaysconcretetechnology.com/concrete-crack-repair-methods.html

Please refer also to Appendix B, for a list of Some Codes, Guidelines, Manuals, Documents, and Books on Assessment, Conservation, Evaluation, Inspection, Maintenance, Preservation, Rehabilitation, Repair, Retrofit, Strengthening & Upgrading Structural Performance.

Appendix

D Guidelines on Selection of Rehabilitation, Repair, Retrofit, Strengthening and Upgrading Methods

This Appendix includes examples of guidelines on selection of repair methods:

Appendix D-1:	Types of remedial action.
Appendix D-2:	Stages and activities during entire life of railway bridges.
Appendix D-3:	Example of decision tree for repair of bridges.
Appendix D-4:	Additional examples of figures, flowcharts, and tables selection of methods of repair/rehabilitation/retrofit.
Appendix D-5:	Relationship between performance verification indices and retrofitting methods (a & b).
Appendix D-6a:	Contents and structure of the new EN 1504.
Appendix D-6b:	Phases of typical repair projects.
Appendix D-6c:	Principles and methods for protection and repair of concrete structures.
Appendix D-7:	Global quality index to determine the rehabilitation decision.
Appendix D-8:	Decisions for structural intervention for retrofitting of building.

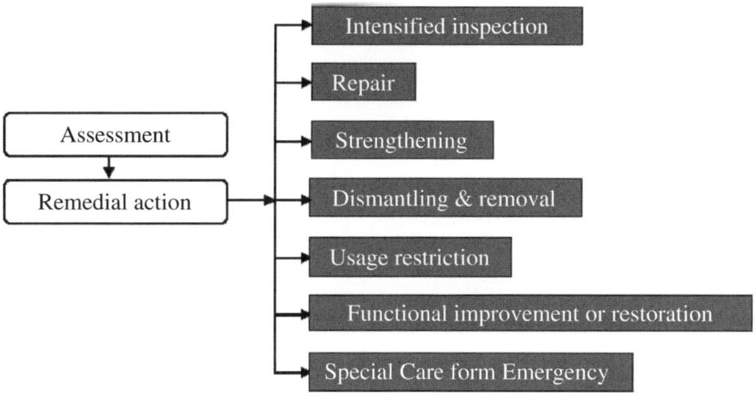

Appendix D-1: Types of remedial action
(From: Concept of Maintenance Part in the JSCE Standard Specifications for Concrete Structures & Its Future Strategy, by Koji Takewaka & Toyoaki Miyagawa, IABSE – fib Conference: Codes in Structural Engineering, Developments & Needs for International Practice, Croatia, May 3–5, 2010, Figure 5)

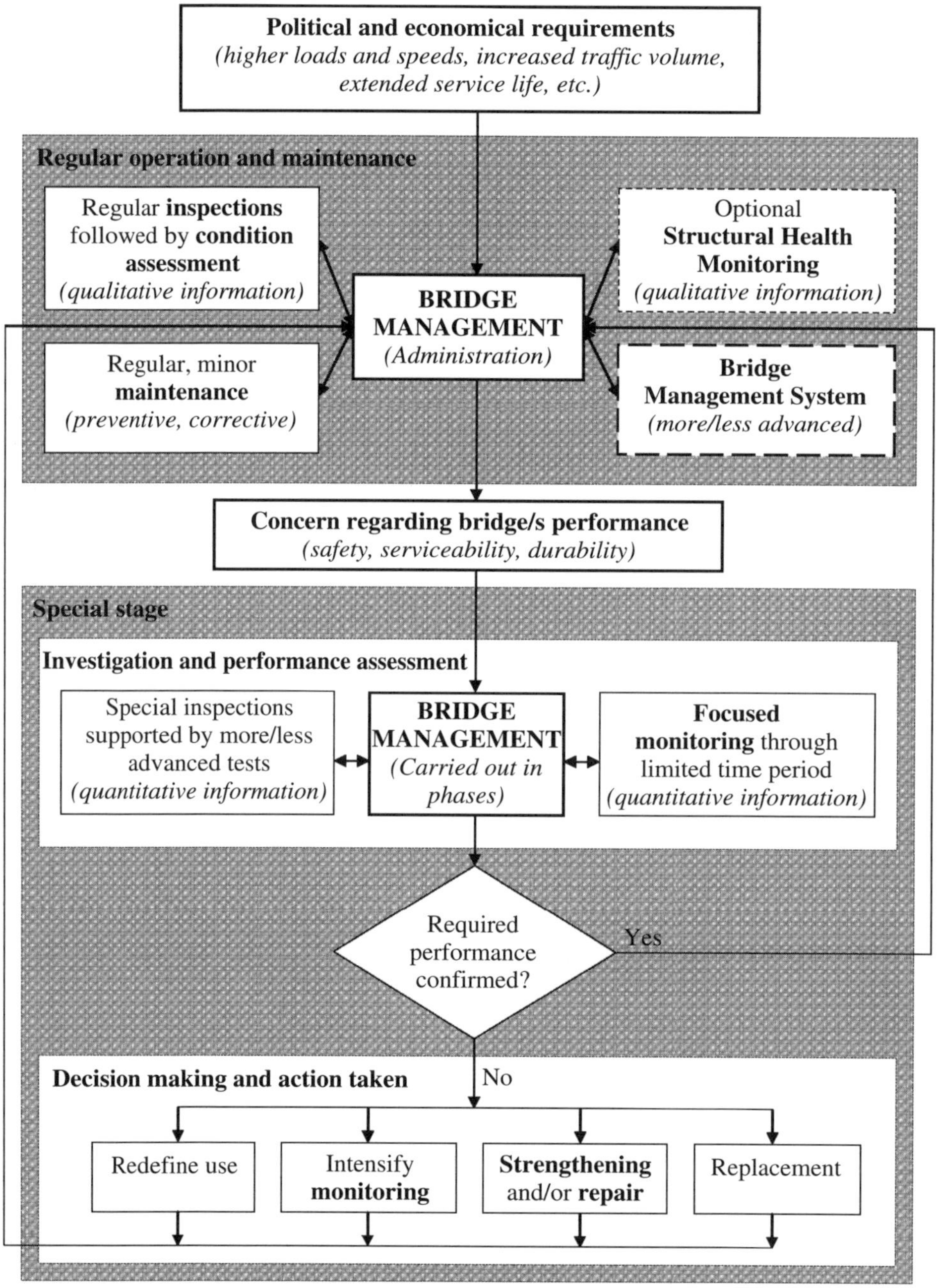

Appendix D-2: Stages and activities during entire life of railway bridges
(Based on: European Guideline for Assessment of Existing Railway Bridges, by J. S. Jensen, J. R. Casas, D. F. Wisniewski, 17th Congress of IABSE, Creating and Renewing Urban Structures, Chicago, USA, September 2008, Figure 2)

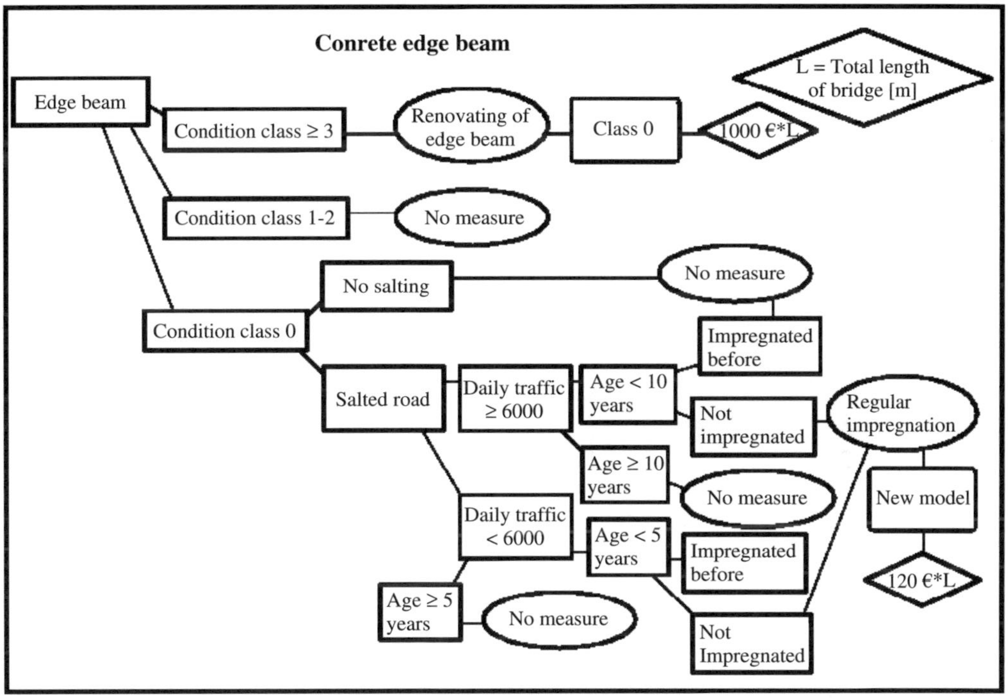

Appendix D-3: Example of decision tree for repair of bridges
(Based on: Anticipation of Repair Need of Bridges Based on: Visual Inspections of Finnish Bridges, by M. Äijälä & J. Lahdensivu, IABSE Symposium, Weimar, Germany, September, 2007)

(1) ***Flowchart for Rehabilitation Actions*** *(Fig. 4.1). Also, Figs. 4.3 to 4.6. Table 4.1*
From: Study on Rehabilitation Actions on Concrete Bridges, Technical Committee on Road Bridges and other Structures (C11), PIARC, World Road Association, 2005,
http://publications.piarc.org/ressources/publications_files/1/629-11-15-VCD.pdf

(2) ***Decision Matrix for Selecting Crack Repair Method*** *(Appendix 2.F, Table 2.F-1)*

(3) ***Substructure Rehabilitation*** *(Appendix 2.D, Flowchart: Fig. 2.D-1). From: Structure Rehabilitation Manual, Policy, Planning and Standards Division Engineering, Ontario, Canada, 2004.* https://www.publications.serviceontario.ca/ecom/MasterServlet/SearchHandler

(4) ***Additional Analysis Required for Existing Bridges Found in Planned Projects*** *(Section 45: Seismic Design & Retrofit, Page: 1.45-5), NJDOT Design Manual for Bridges & Structures,* http://www.state.nj.us/transportation/eng/documents/BDMM/pdf/bmsec45.pdf

(5) ***Seismic retrofitting process for highway bridges*** *(Figs. 1–10, Page 27). From: Seismic Retrofitting Manual For Highway Structures: Part 1—Bridges, by I. Buckle, I. Friedland, J. Mander, G. Martin, R. Nutt, M. Power, FHWA-HRT-06-032, 2006.*
http://mceer.buffalo.edu/publications/Bridge_and_Highway_Reports/seismic_retrofit_manual/default.asp

(6) ***Class diagram*** *(Fig. 4, page 23)*
From: Multi-Objective Optimization for Bridge Management Systems, by V. Patidar, S. Labi, K. C. Sinha, NCHRP Report 590, Transportation Research Board, USA, 2007.

Appendix D-4: Additional examples of figures, flowcharts and tables selection of methods of repair/rehabilitation/retrofit

Appendix D5a: Relationship between Performance Verification Indices & Retrofitting Methods (From: JSCE_Guidelines for Retrofit of Concrete Structures–Draft, Table S7.1.1 (a))Translation from the CONCRETE LIBRARY No.95 published by JSCE, September 1999, JSCE Working Group on Retrofit Design of Concrete Structures in Specification Revision Committee http://www.jsce.or.jp/committee/concrete/e/newsletter/newsletter01/recommendation/FRP-sheet/3.pdf

Retrofitting Objective	Measure	Specific Construction Method Name*1	Member*2								
			All	Beam	Column	Slab	Wall*3	Bearing	Pile	Foundation	Other
Concrete member retrofitting	Bonding	Steel plate bonding method (bonding method)		•	O	•	O				
	Jacketing	Continuous fiber sheet bonding method (bonding method)		•	O	•	O				
		Continuous fiber retrofitted plate bonding method		O	O	O	O				
		Steel plate jacketing method (jacketing method)			•		O				
		Continuous fiber sheet jacketing method (jacketing method)			•		O				
		Continuous fiber retrofitted plate jacketing method			O						
		Retrofitted concrete jacketing method (jacketing method)			•		O				
		Precast concrete jacketing method			•		O				
		Mortar spray method (jacketing method)			•		O				
		Precast panel jacketing method (Jacketing method)			•						
	Introduction of prestressing	External cable method (external cable method)		•	O	O					
		Prestressing introduction (internal cable) method		•	O		O				
	Overlaying for Sections	Upper surface overlaying method (overlaying method)		O		•					
		Lower surface overlaying method (overlaying method)		O		•					
		Lower surface spray method (overlaying method)		O		•					
	Member replacement	Repaving method		O	O	•	•				
Retrofitting as a structural body	Addition of beams (girders)	Beam (girder) addition method		•		•					
	Addition of walls	Seismic wall addition method					•				
	Addition of support points	Support point addition method		•		•					
	Seismic isolations	Seismic isolation method	•					•			
Foundation retrofitting	Addition of walls and beams	Underground wall (beam) addition method		O			O				•
	Addition of piles and footing	Pile (footing) addition method							•		•
		Foundation improvement method								•	•
		Steel sheet pile coffering method								•	•
	Improvement of foundation	Foundation compacting method								•	•
	Ground anchors	Ground anchor method					O				•
Bearing retrofitting and prevention of bridge collapse	Replacement of bearings (including seismic isolations)	Bearing replacement method	•					•			
	Retrofitting of bearings themselves	Bearing retrofitting method						•			
	Retrofitting of bearings and bridge seats	Shoe/Seat retrofitting method		O	O			•			
	Movement restrictions	Movement restrictions equipment installation method		O	O			•			
	Shift to continuous girders	Main girder connection method		•				•			
	Gerber girders	Balance method		•				•			
	External cables	Elasticity restraint method		•							
Repair of cracks and missing sections	Injection	Crack fill method		•	•	•					
	Fill	Fill method		•	•	•					
	Repair	Section repair method		•	•	•					
Ensuring Transit routes on bridge road surface	Repairs to bridge surface	Pavement repair method									•
		Expansion joint repair method									•

Note:
*1 Parentheses after method names indicate correspondence with method names noted in these (draft) guidelines
*2 •: Used comparatively often O: Use is thought to be possible
*4 •: Thought to be very effective O: Somewhat effective but effect is thought to be comparatively minor (indirect)
*3 Including wall type bridge piers
*5 Effective primarily for punching shear

Appendix D-5b: Relationship between Performance Verification Indices & Retrofitting Methods. (From: JSCE_Guidelines for Retrofit of Concrete Structures - Draft, Table S7.1.1 (b)) Translation from the CONCRETE LIBRARY No.95 published by JSCE, September 1999, JSCE Working Group on Retrofit Design of Concrete Structures in Specification Revision Committee. http://www.jsce.or.jp/committee/concrete/e/newsletter/newsletter01/recommendation/FRP-sheet/3

Retrofitting Objective	Measure	Specific Construction Method Name*1	Performance Verification Index *4										
			Safety							Serviceability			Restorability
	Bonding	Steel plate bonding method (bonding method)	Axial force	Bending	Shear	Torsion	Fatigue	Rigid body stability	Durability	Crack width / density	Deflection displacement	Other	
Concrete member retrofitting	Jacketing	Continuous fiber sheet bonding method (bonding method)	●	●	●	●	●			●	●		
		Continuous fiber retrofitted plate bonding method	○	●	●	●	●			●	○		
		Steel plate jacketing method (jacketing method)	○	●	●	●	●			●	○		
		Continuous fiber sheet jacketing method (jacketing method)	●	●	●	●			●	●	●		●
		Continuous fiber retrofitted plate jacketing method	●	●	●	●			●	○	○		●
		Retrofitted concrete jacketing method (jacketing method)	●	●	●	●			○	○	○		○
		Precast concrete jacketing method	●	●	●	●			●	●	●		●
		Mortar spray method (jacketing method)	●	●	●	●			●	●	●		●
		Precast panel jacketing method (Jacketing method)	●	●	●	●			●	●	●		●
	Introduction of prestressing	External cable method (external cable method)	●	●	●	●			●	●	●		●
		Prestressing introduction (internal cable) method	○	●	●	●	●			●	●		
	Overlaying for Sections	Upper surface overlaying method (overlaying method)	○	●	●	●	●		○	●	●		
		Lower surface overlaying method (overlaying method)	○	●	●*5	●	●			●	●		
		Lower surface spray method (overlaying method)	○	●	○*5	●	●			●	●		
	Member replacement	Repaving method	○	●	○*5	●	●			●	●		
Retrofitting as a structural body	Addition of beams (girders)	Beam (girder) addition method	●	●	●	●	●		○	●	●		○
	Addition of walls	Seismic wall addition method		●	●	●	●			●	●		
	Addition of support points	Support point addition method		●	●	●					●		
	Seismic isolations	Seismic isolation method		●	●	●					●		
Foundation retrofitting	Addition of walls and beams	Underground wall (beam) addition method		●	●	●					●		
	Addition of piles and footing	Pile (footing) addition method	○	●	●	●		●		○	○		
		Foundation improvement method	○	●	●	●		●		○	○		
		Steel sheet pile coffering method						●			○		
	Improvement of foundation	Foundation compacting method						●			○		
	Ground anchors	Ground anchor method						●			○		
Bearing retrofitting and prevention of bridge collapse	Replacement of bearings (including seismic isolations)	Bearing replacement method						●			○		
	Retrofitting of bearings themselves	Bearing retrofitting method									○	●	○
	Retrofitting of bearings and bridge seats	Shoe/Seat retrofitting method									○	●	○
	Movement restrictions	Movement restrictions equipment installation method									○	●	○
	Shift to continuous girders	Main girder connection method									○	●	○
	Gerber girders	Balance method									○	●	○
	External cables	Elasticity restraint method									○	●	
Repair of cracks and missing sections	Injection	Crack fill method									○	●	
	Fill	Fill method			○	○	○			●	○	●	
	Repair	Section repair method	○	○	○	○	○			●	○	●	
Ensuring Transit routes on bridge road surface	Repairs to bridge surface	Pavement repair method	○	○	○	○	○			●		●	
		Expansion joint repair method										●	

Note:

*1 Parentheses after method names indicate correspondence with method names noted in these (draft) guidelines

*2 ●: Used comparatively often　　O: Use is thought to be possible　　*3 Including wall type bridge piers

*4 ●: Thought to be very effective　　O: Somewhat effective but effect is thought to be comparatively minor (indirect)　　*5 Effective primarily for punching shear

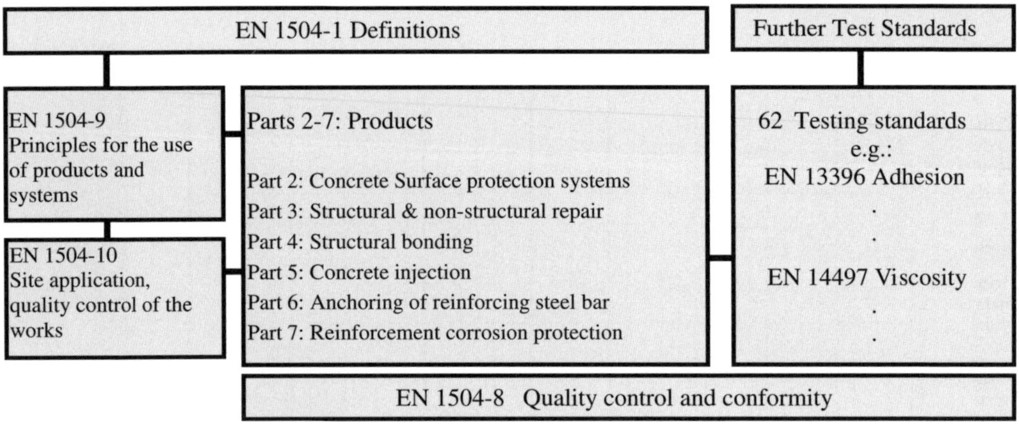

Appendix D-6a: Contents and structure of the new EN 1504
(Based on: EN 1504 Products and systems for the protection and repair of concrete structures—Definitions, requirements, qualify control and evaluation of conformity)

Project Phases	**Basic considerations and actions** & Relevant Clause in EN 1504
Information about the structure (1)	• Condition and history of structure • Documentation • Previous repair and maintenance
Process of assessment (1)	• Defects and their classification and causes • Safety / structural appraisal before protection and repair
Management strategy (2) & (3)	• Options • Principles • Methods • Safety / structural appraisal during protection & repair
Design of repair work (3), (4) & (6)	• Intended use of products • Requirements: substrate,Products ,Work • Specifications • Drawings • Safety / structural appraisal after protection and repair
Repair work (3), (4), (6), (7)	• Choice and use of products and systems and methods and equipment to be used • Tests of quality control • Health and safety
Acceptance of repair work (5)	• Acceptance testing • Remedial works • Documentation

(1). Clause 4: Minimum requirements before protection and repair
(2). Clause 5: Protection & Repair within a structure management strategy
(3). Clause 6: Basis for the choice of protection and repair principles and methods
(4). Clause 7: properties of products and systems required for compliance with the principles of protection & repair
(5). Clause 8: Maintenance of following completion of protection and repair
(6). Clause 9: Health, safety & the environment
(7). Clause 10: Competence of personnel

Appendix D-6b: Phases of typical repair projects
(Based on: EN 1504-9: Products and systems for the protection and repair of concrete structures. Principles for the use of products and systems, Figure A1)

Principle and its Definition	Examples of Methods of Protection or Repair Based on the Principles	Relevant part of EN 1504
Principle 1 [PI] Protection against ingress Reducing or preventing the ingress of adverse agents, e.g. water, other liquids, vapour, gas, chemicals and biological agents.	1.1 Hydrophobic impregnation	2
	1.2 Impregnation	2
	1.3 Coating	2
	1.4 Surface bandaging of cracks	
	1.5 Filling of cracks	5
	1.6 Transferring cracks into joints	
	1.7 Erecting external panels	
	1.8 Applying membranes	
Principle 2 [MC] Moisture Control Adjusting and maintaining the moisture content in the concrete within a specified range of values	2.1 Hydrophobic impregnation	2
	2.2 Impregnation	2
	2.3 Coating	2
	2.4 Erecting external panels	
	2.5 Electrochemical treatment	
Principle 3 [CR] Concrete Restoration Restoring the original concrete of an element of the structure to the originally specified shape and function. Restoring the concrete structure by replacing part of it.	3.1 Hand-applied mortar	3
	3.2 Recasting with concrete or mortar	3
	3.3 Spraying concrete or mortar	3
	3.4 Replacing elements	
Principle 4 [SS] Structural Strengthening Increasing or restoring the structural load bearing capacity of an element of the concrete structure.	4.1 Adding or replacing embedded or external reinforcing bars	6
	4.2 Adding reinforcement anchored in pre-formed or drilled holes	
	4.3 Bonding plate reinforcement	4
	4.4 Adding mortar or concrete	3, 4
	4.5 Injecting cracks, voids or interstice	5
	4.6 Filling cracks, voids or interstices	5
	4.7 Prestressing (post tensioning)	
Principle 5 [PR] Physical resistance Increasing resistance to physical or mechanical attack	5.1 Coating	2
	5.2 Impregnation	2
	5.3 Adding mortar or concrete	3
Principle 6 [RC] Resistance to chemicals Increasing resistance of the concrete surface to deteriorations by chemical attack.	6.1 Coating	2
	6.2 Impregnation	2
	6.3 Adding mortar or concrete	3
Principle 7 [RP] Preserving or Restoring Passivity Creating chemical conditions in which the surface of the reinforcement is maintained in or is returned to a passive condition.	7.1 Increasing cover with additional mortar or concrete	3
	7.2 Replacing contaminated or carbonated concrete	3
	7.3 Electrochemical realkalisation of carbonated concrete	
	7.4 Realkalisation of carbonated concrete by diffusion	
	7.5 Electrochemical chloride extraction	
Principle 8 [IR] Increasing Resistively Increasing the electrical resistivity of the concrete.	8.1 Hydrophobic impregnation	2
	8.2 Impregnation	2
	8.3 Coating	2
Principle 9 [CC] Cathodic Control Creating conditions in which potentially cathodic areas of reinforcement are unable to drive an anodic reaction.	9.1 Limiting oxygen content (at the cathode) by saturation or surface coating	
Principle 10 [CP] Cathodic Protection	10.1 Limiting oxygen content (at the cathode) by saturation or surface coating	
Principle 11 [CA] Control of Anodic areas Creating conditions in which potentially anodic areas of reinforcement are unable to take part in the corrosion	11.1 Active coating of the reinforcement	**7**
	11.2 Barrier coating of the reinforcement	**7**
	11.3 Applying corrosion inhibitors in or to the concrete	

Appendix D-6c: Principles and Methods for protection and repair of concrete structures (Based on: EN 1504-9: Products and systems for the protection and repair of concrete structures. Principles for the use of products and systems, Table 1)

Class of technical state	Value of the global index	General appreciation of the technical state	Required measures Rehabilitation decision
I	80 - 100	Very good state. Small local defects	Local repairs, preservation
II	70 - 80	Good state. Small defects on limited areas	Repairs of affected areas
III	60 - 70	Satisfactory state. Defects of important extent, but without fundamentally affecting the structure	General repairs and partial consolidations
IV	50 - 60	Unsatisfactory state. Generalized defects essentially affecting the structure	Capital repairs, partial or general consolidation
V	< 50	Unsatisfactory state. The minimum safety and functionality are not assured	General consolidation, partial or total demolition

$$I_{st} = \sum_{j=1}^{m} P_{j_t} \cdot I_{st}, \quad \text{where} \sum_{j=1}^{m} P_j = 1 \quad I_{st} = k \cdot \sum_{j=1}^{10} C_i \quad C_i = 10 - D$$

Appendix D-7: Global quality index to determine the rehabilitation decision (Based on: Rehabilitation of Reinforced Concrete Silos, by I. Bucur-Horvath, I. Popa, A. Hedes, IABSE Symposium, Responding to Tomorrow's Challenges in Structural Engineering, Budapest, September 2006)

Type of intervention for Retrofitting of Structural Elements

a) Local or overall modification of damaged or undamaged elements (repair, strengthening or full replacement), considering the stiffness, strength and/or ductility of these elements.
b) Addition of new structural elements (e.g. bracings or infill walls; steel, timber or reinforced concrete belts in masonry construction; etc.).
c) Modification of the structural system (elimination of some structural joints; widening of joints; elimination of vulnerable elements; modification into more regular and/or more ductile arrangements)
d) Addition of a new structural system to sustain some or all of the entire seismic action.
e) Possible transformation of existing non-structural elements into structural elements.
f) Introduction of passive protection devices through dissipative bracing or base isolation.
g) Mass reduction.
h) Restriction or change of use of the building.
i) Partial demolition.

(1) An intervention may be selected from the indicative types mentioned above.

(2) One or more types in combination may be selected. In all cases, the effect of structural modifications on the foundation should be taken into account.

(P) The selection of the type, technique, extent and urgency of the intervention shall be based on the structural information collected during the assessment of the building.

Appendix D-8: Decisions for Structural Intervention for Retrofitting of Building (Based on: EN 1998-3:2005 Eurocode 8. Design of Structures for Earthquake Resistance. Part 3: Assessment and Retrofitting of Buildings (2005), Section 5.1.2)